Mathematics

Irina Amlin
Rita Bateson
Series editor: Paul Morris

Acknowledgements

From Rita Bateson: With thanks to my first and most enthusiastic maths teacher, my dad, Sean, and to my ever-supportive mother, Eileen, as well as to my wonderful siblings. Special thanks to my proof-reading brother, Sean. A million thanks to Caroline, Mary, Maria and Patricia for their enduring friendship and to Robert for his unfailing kindness and wisdom.

Thanks also to the incredible staff at Hodder Education, especially So-Shan, Megan and the meticulous Anna. To another wonderful teacher, Mrs Hainsworth, and all the inspiring educators and students I have worked with over the years. My thanks and gratitude especially to my co-author, Irina.

Most of all, this book is dedicated to my incredible daughter Ellie and amazing, patient and kind husband Andrew, without whom nothing would ever get done.

From Irina Amlin: This book is dedicated to my two remarkable daughters, Penelope and Cora, who I know will each achieve great things in their own right. Thank you for giving up so much playtime with mummy to allow her to work on this!

Thank you also to my husband Ryan, for his support and encouragement. My heartfelt gratitude goes to my mother Dara and brother Luka, who taught me how to learn, and to my father and chief adviser Janko, who taught me how to teach! To Fred Ferneyhough, who guided my decision to pursue teaching. To my many inspiring colleagues, especially Rita Bateson, Stephen Pulford and Graham Lewis – our invigorating discussions and joint projects have been my best PD. Thank you to Dr Jim Christopher, Charlie Judd and Carlos Symonds for encouraging creativity in mathematics. Finally, thank you to So-Shan, Megan, and all the patient, hard-working staff at Hodder Education, as well as to Dr Anna Clark for your keen eye!

Although every effort has been made to ensure that website addresses are correct at time of going to press, Hodder Education cannot be held responsible for the content of any website mentioned in this book. It is sometimes possible to find a relocated web page by typing in the address of the home page for a website in the URL window of your browser.

Hachette UK's policy is to use papers that are natural, renewable and recyclable products and made from wood grown in sustainable forests. The logging and manufacturing processes are expected to conform to the environmental regulations of the country of origin.

Orders: please contact Bookpoint Ltd, 130 Milton Park, Abingdon, Oxon OX14 4SB. Telephone: (44) 01235 827720. Fax: (44) 01235 400454. Lines are open from 9.00–5.00, Monday to Saturday, with a 24 hour message answering service. You can also order through our website www.hoddereducation.com

Cover photo © claporte/iStockphoto/Getty Images
Illustrations by DC Graphic Design Limited & Richard Duszczak
Typeset in Frutiger LT Std 45 Light 11/15pt by DC Graphic Design Limited, Hextable, Kent
Printed in India

A catalogue record for this title is available from the British Library

ISBN 9781471881039

Contents

How to use this book

Welcome to Hodder Education's *MYP by Concept* series! Each chapter is designed to lead you through an *inquiry* into the concepts of mathematics, and how they interact in real-life global contexts.

The *Statement of Inquiry* provides the framework for this inquiry, and the *Inquiry questions* then lead you through the exploration as they are developed through each chapter.

KEY WORDS

Key words are included to give you access vocabulary for the topic. **Glossary** terms are highlighted and, where applicable, search terms are given to encourage independent learning and research skills.

As you explore, activities suggest ways to learn through *action*.

■ ATL

■ Activities are designed to develop your *Approaches to Learning* (ATL) skills.

Detailed information or explanation of certain points is given whenever necessary. Key Approaches to Learning skills for MYP Mathematics are highlighted whenever we encounter them.

Hint

In some of the Activities, we provide Hints to help you work on the assignment. This also introduces you to the new Hint feature in the on-screen assessment in MYP5.

Each chapter is framed with a *Key concept* and a *Related concept,* and is set in a *Global context.*

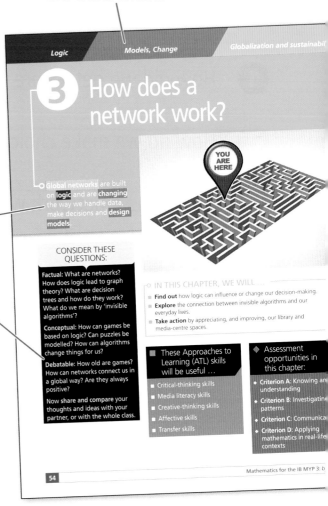

Logic | Models, Change | Globalization and sustainabil

③ How does a network work?

YOU ARE HERE

Global networks are built on logic and are changing the way we handle data, make decisions and design models.

CONSIDER THESE QUESTIONS:

Factual: What are networks? How does logic lead to graph theory? What are decision trees and how do they work? What do we mean by 'invisible algorithms'?

Conceptual: How can games be based on logic? Can puzzles be modelled? How can algorithms change things for us?

Debatable: How old are games? How can networks connect us in a global way? Are they always positive?

Now share and compare your thoughts and ideas with your partner, or with the whole class.

○ IN THIS CHAPTER, WE WILL ...
■ **Find out** how logic can influence or change our decision-making.
■ **Explore** the connection between invisible algorithms and our everyday lives.
■ **Take action** by appreciating, and improving, our library and media-centre spaces.

■ These Approaches to Learning (ATL) skills will be useful ...
■ Critical-thinking skills
■ Media literacy skills
■ Creative-thinking skills
■ Affective skills
■ Transfer skills

◆ Assessment opportunities in this chapter:
◆ **Criterion A:** Knowing and understanding
◆ **Criterion B:** Investigating patterns
◆ **Criterion C:** Communica
◆ **Criterion D:** Applying mathematics in real-life contexts

◆ Assessment opportunities in this chapter:

◆ Some activities are *formative* as they allow you to practise certain of the MYP Mathematics *Assessment Criteria*. Other activities can be used by you or your teachers to assess your achievement against all strands of an Assessment Criterion.

▼ Links to:

Like any other subject, mathematics is just one part of our bigger picture of the world. Links to other subjects are discussed.

Each chapter covers one of the four branches of mathematics identified in the MYP Mathematics skills framework.

We have incorporated Visible Thinking – ideas, framework, protocol and thinking routines – from Project Zero at the Harvard Graduate School of Education into many of our activities. You are prompted to consider your conceptual understanding in a variety of activities throughout each chapter.

Finally, at the end of each chapter, you are asked to reflect back on what you have learned with our Reflection table, maybe to think of new questions brought to light by your learning.

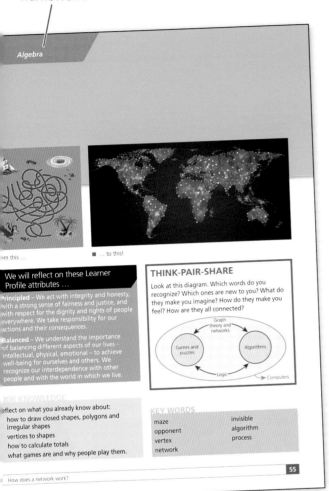

Reflection

Use this table to reflect on your own learning in this chapter.						
Questions we asked	Answers we found	Any further questions now?				
Factual						
Conceptual						
Debatable						
Approaches to Learning you used in this chapter:	Description – what new skills did you learn?	How well did you master the skills?				
		Novice	Learner	Practitioner	Expert	
Learner Profile attribute(s)	Reflect on the importance of this attribute for your learning in this chapter.					

! Take action

! While the book provides opportunities for action and plenty of content to enrich the conceptual relationships, you must be an active part of this process. Guidance is given to help you with your own research, including how to carry out research, how to make changes in the world informed by mathematics, and how to link and develop your study of mathematics to the global issues in our twenty-first century world.

Worked examples and Practice questions are given in colour-coded boxes to show the level of difficulty:

Problem   **Complex**

Challenging   **Unfamiliar**

1 How did we get where we are?

Firstly humans observed phenomena and relationships. Then they **measured** quantities. Soon they could create **general rules and formulae** which could be **justified**. All these ways of knowing come together to give us our **mathematical body of knowledge**.

CONSIDER THESE QUESTIONS:

Factual: How do we construct a number line and why? How do we operate with numbers? What's new in mathematics?

Conceptual: How did numbers take form and shape? Why do we find rules and then try to justify them? How are measurements 'born' and why do they die?

Debatable: What came after counting? How do our surroundings affect our identities as mathematicians, and vice versa? How can we use mathematics to reveal past or predict future behaviours? Is mathematics discovered or invented? Is a more commonly used unit the best one to go with?

Now **share and compare** your thoughts and ideas with your partner, or with the whole class.

IN THIS CHAPTER, WE WILL...

- **Find out** about individuals and societies that contributed to the body of knowledge that is mathematics.
- **Explore** what drove mathematical discoveries and inventions such as the equals sign, negative numbers, algebra or fractals.
- **Take action** by contributing to the discovery of new prime numbers.

These Approaches to Learning (ATL) skills will be useful …

- Creative-thinking skills
- Critical-thinking skills
- Information literacy skills
- Communication skills

We will reflect on this Learner Profile attribute …

- **Communicator** – We express ourselves confidently and creatively in more than one language and in many ways. We collaborate effectively, listening carefully to the perspectives of other individuals in the group.

Assessment opportunities in this chapter:

- **Criterion A:** Knowing and understanding
- **Criterion B:** Investigating patterns
- **Criterion C:** Communicating
- **Criterion D:** Applying mathematics in real-life contexts

THINK-PAIR-SHARE

Why do we write 'mathematics' as a plural but refer to it as a singular object? We never say 'My mathematics are fun today', or if asked 'When do you have mathematics?' we don't say 'I have *them* today'. Why do you think this is? How might this situation have developed?

Discuss your thoughts with a partner and then share with the rest of the class.

PRIOR KNOWLEDGE

Reflect on what you already know about:
- the ancient subject of mathematics
- how to count up to large numbers and how to represent them using mathematical symbols (numbers)
- prime numbers, decimals, fractions and integers
- how to operate (add, subtract, multiply, divide and square) with numbers
- the ancient civilizations of the Egyptians, Babylonians, Chinese and Mayans, among others.

KEY WORDS

rotation	correspondence
culture	Imperial
society	metric

How did numbers take form and shape?

'We live in a society exquisitely dependent on science and mathematics, in which hardly anyone knows anything about science and mathematics. This is a clear prescription for disaster.'

Carl Sagan, 1990

HOW DID WE GET WHERE WE ARE?

Mathematics is a wonder, a tool, a queen of knowledge. Throughout time humans have formed their mathematical knowledge and discoveries by observing and measuring the world around them. As an area of knowledge, mathematics grew hand in hand with civilization itself. In fact, many advances were made possible by discoveries in mathematics – pyramids, money, bridges and computers are all examples of this.

Informal mathematics has been used for much longer than the formal version: hunters used ideas of 'likely' or 'unlikely' without calculating actual probabilities; farmers identified patterns in crop growth, weather and animal breeding; families had to share resources such as food and space using proportional reasoning.

Knowledge is gained and lost through time. We know from our own lives that we learn and we also forget things. Mathematics throughout history is the same. Inventions and discoveries have occurred at different times and in different places, then sometimes forgotten forever or rediscovered centuries later. How can this happen?

Newgrange is a tomb in Ireland from the Stone Age, which was built over 5000 years ago. Read the extract below to learn more about it.

www.newgrange.com

Above the entry to the tomb, there is a small opening called a roof-box. Its purpose is to allow sunlight to travel through a 19-metre long passage into a central chamber on the shortest days of the year, around December 21st, the winter solstice. At dawn, from December 19th to 23rd, a narrow beam of light penetrates the roof-box and reaches the floor of the chamber, gradually extending to the rear of the chamber.

As the Sun rises higher, the beam widens within the chamber so that the whole room becomes dramatically illuminated. This event lasts for 17 minutes, beginning around 9am. The accuracy of Newgrange as a time-telling device is remarkable when one considers that it was built 500 years before the Great Pyramids and more than 1000 years before Stonehenge.

Source: **www.newgrange.com**

It is clear that the people who built Newgrange understood a great deal about the Sun, angles and time, but we have no record of what or how they knew these things. The nature of the writing materials in ancient times may play a role here. Carvings that were marked in clay survived better than words written on scrolls or papyrus. Hieroglyphics inside pyramids stood the test of time far better than pictographs painted in caves or on cliffsides.

Key developments in mathematics occur when relationships are turned into general rules and formulae, which are then tested and justified. There is a huge leap from **utility** (usefulness) into **abstraction** (moving from specific events or examples into general rules). People have been thinking about numbers and mathematical ideas for thousands of years, over a hundred generations.

ACTIVITY: In what order did these events happen?

■ ATL

- Creative-thinking skills: Use brainstorming and visual diagrams to generate new ideas and inquiries

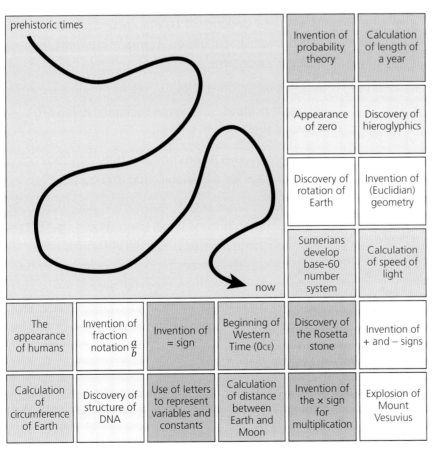

prehistoric times

Invention of probability theory	Calculation of length of a year
Appearance of zero	Discovery of hieroglyphics
Discovery of rotation of Earth	Invention of (Euclidian) geometry
Sumerians develop base-60 number system	Calculation of speed of light

now

The appearance of humans	Invention of fraction notation $\frac{a}{b}$	Invention of = sign	Beginning of Western Time (0$_{CE}$)	Discovery of the Rosetta stone	Invention of + and − signs
Calculation of circumference of Earth	Discovery of structure of DNA	Use of letters to represent variables and constants	Calculation of distance between Earth and Moon	Invention of the × sign for multiplication	Explosion of Mount Vesuvius

Each of the small boxes in the diagram holds a moment in humankind's understanding or use of mathematics. Working in pairs or small groups, decide the order in which they happened. Discuss why you think some happened before others. When you have agreed an order for all 20 events, your teacher will share the answers with you. Reflect on how close your estimates were to the correct order.

Once you have confirmed the correct order, your next task is to estimate **approximately** when these events happened. Start by predicting the century in which you think an event happened and, if you are feeling confident, you could have a guess at which decade it comes from.

◆ Assessment opportunities

- ◆ In this activity you have practised skills that are assessed using Criterion A: Knowing and understanding, and Criterion C: Communicating.

THE FIRST APPEARANCE OF MATHEMATICS FOR HUMANS

When children are very young, one of the first things they are taught is how to count. They practise the numbers over and over so they can recognize quantity and compare relative sizes.

'People used to think that literacy was a gift from the gods.' What does this statement make you think of? What does it indicate about a particular place or time? When do you think it might have been said?

In *Mathematics for the IB MYP 2*, Chapter 6, we were introduced to a mystery tablet that dates from over 5000 years ago and which could not be **deciphered** (decoded). The mystery of this tablet, and others like it, introduced us to the concept of **correspondence counting**. This type of counting uses a symbol to represent a quantity. How were the symbols formed? What did they mean to the traders, as goods entered and left the marketplace?

These symbols represent a way to track quantities **without** a common numerical form. This form of representation made it possible to add and subtract without using arithmetic. The invention of numbers to represent quantities has occurred several times in different places around the world.

Imagine you lived in a world without numbers. How would you keep track of things? There have been many ingenious ways created by humans over time to solve this problem, and not all led to the invention of numbers. Here are some other interesting counting techniques:

- Finger counting in African tribes, using various positions of fingers to represent different numbers.
- The Yoruba of Kenya used shells for correspondence counting without numbers to identify missing herd members. In the more recent past, a similar technique was used by cricket umpires to track the number of balls in an over.
- The Incas used rope-and-knot systems called quipu to keep track of quantities.

THE DEVELOPMENT OF NUMBERS

The earliest evidence of counting comes from Mesopotamian cuneiform, Egyptian hieroglyphics and Chinese ideograms.

Read more about the development of numbers in different cultures here: **www.archimedes-lab.org/numeral.html**

Number provides a common link between societies and is a basis for communication and trade.

The idea of 'single versus many' must have developed early in human history. It has been shown that babies and even some animals understand fairness and unfairness. Early humans probably understood when someone had more than they did, even if they didn't have the language or symbols (numbers) to express it. Perhaps our mathematical understanding started with these two crucial questions: 'How many?' and 'How much?'

In many cultures, odd numbers developed as 'male' and even numbers were 'female'. Tallying, ordering and counting seem to have originally served very similar purposes.

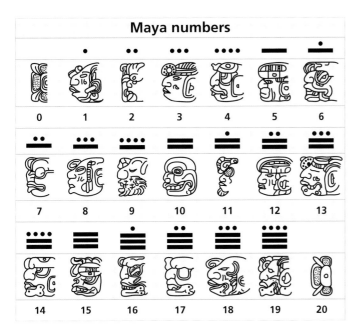

Sumerian cuneiform numbers

1	2	3	4	5	6	7	8	9	10

Brahmi ➡		—	=	≡	+	⋏	⋐	⌐	ꙅ	?
Hindu ➡	०	?	?	३	४	५	६	७	८	९
Arabic ➡	•	١	٢	٣	٤	٥	٦	٧	٨	٩
Medieval ➡	0	1	2	3	𝒳	𝒻	6	ʌ	8	9
Modern	0	1	2	3	4	5	6	7	8	9

Maya numbers

0	1	2	3	4	5	6

7	8	9	10	11	12	13

14	15	16	17	18	19	20

ROMAN NUMERALS

I = 1	VI = 6
II = 2	VII = 7
III = 3	VIII = 8
IV = 4	IX = 9
V = 5	

Using Roman numerals, all numbers can be written with combinations of only seven signs.

I = 1
V = 5
X = 10
L = 50
C = 100
D = 500
M = 1000

PRACTICE EXERCISE

Convert the following numbers to Roman numerals.

1	7	3	700	5	89
2	17	4	51	6	90

All numbers were made from these signs, combining them to make different values and always writing the symbols in descending order. A number like 52 would be made from L (50) plus I plus I = LII.

For example:

MDCCLXV = 1765

VIII = 8

If a smaller number is written to the left of a larger number, then the smaller number is taken away from the next, larger number. For example, the number 9 is written as IX because this signifies 1 subtracted from 10. And XIX shows a value of 19 (10 plus [1 subtracted from 10]). Another rule is that no more than three similar symbols can be used.

We still see traces of Roman numerals around us today:
- Sometimes the year a building was built is etched into the stone in Roman numerals.
- Until recently, many TV shows used Roman numerals in the end credits to indicate the year; BBC shows still do this.
- Royalty and popes have Roman numerals after their names to show their order: Pope John Paul II, King Henry VII. This also occurs when several generations of the same family share a name, for example William Morrison III.
- The quadrants in Cartesian geometry are labelled using these numbers.
- Competitions and championships are often numbered using Roman numerals, including the Olympic Games (XXIX was in Beijing) and Super Bowl (such as XXXIX).

Here are some other examples. How are Roman numerals being used in these situations?

- Apple iPhone X
- Super Bowl LII
- Meeting of the Minds IX

What came after counting?

STATISTICS OR STATE-ISTICS

Administrators and rulers have always needed to calculate taxes and count money. It is a very small step from **counting** to **accounting**. Accounting is keeping records of individual or business financial affairs. If one of the original reasons for counting was to see how much of something you owned, had or sold, then it makes sense that you might also want to record that information. It would help you keep track of your livestock (animals) or your money. In fact, in many civilizations, keeping track of such numbers was the driving force for the development of mathematics.

We often think of statistics as a relatively modern invention, but we can see the beginnings of this branch of mathematics in history. Statistics is defined as dealing with the **collection, classification, analysis, and interpretation of numerical facts or data**. We know that the collection and analysis of data began long ago, but in an informal way. You will see more about the beginning of formal statistics later in this chapter.

In ancient Uruk, at the market place, Ali is a trader looking to make exchanges. He arrives with 2000 heads of corn, 50 squash and three pigs. He trades two pigs for another 1000 heads of corn and then sells 1500 heads of corn and 30 squash for gold.

Why is counting important here? What increases or decreases have occurred? Can you find any other mathematical ideas in these transactions? Discuss your ideas as a class.

WHAT IS LESS THAN 1?

The way we represent fractions today came from the Hindu system, Brahmi, which had nine symbols and a zero. It came into more general use through the Arab mathematicians, who themselves invented the horizontal bar separating the numerator and denominator around 1200CE. Fibonacci was the first European mathematician to use a fraction bar as it is used today. Muslim mathematicians were the first to use decimal numbers on a large scale instead of fractions from the 15th century, but Simon Stevin (1548–1620CE) established their regular use and he advocated that decimalization should happen in money, weights and time.

1800 BCE
Egyptians using fractions

Nope, I don't get it. Leave it to the Swedes. They will invent bar charts in 2500 years!

Did you know?

The word **fraction** comes from the Latin word 'fractio', which means 'to break'.

1650

Invention of fraction notation

$$\frac{a}{b}$$

ACTIVITY: The introduction of fractions

■ ATL

■ Communication skills: Organize and depict information logically

The Egyptians were the first to use fractions, as early as 1800BCE.

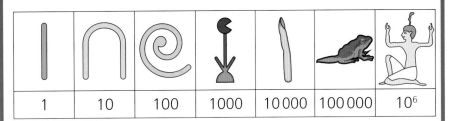

1	10	100	1000	10 000	100 000	10^6

The hieroglyphs show the images the Egyptians used for the numbers we know now.

The Egyptians wrote fractions using an eye symbol to mean 'one part of'. This symbol was used in combination with a number below to show the denominator.

One-fifth or $\frac{1}{5}$ was written like this:

1 **Express the following images as fractions.**

2 **Now create hieroglyphs to represent these fractions.**
   a $\frac{1}{100}$ b $\frac{1}{15}$ c $\frac{1}{22}$

The Egyptians only ever used a symbol on the top line, not any other number. This meant they could only use unit fractions (fractions with 1 on the top line) and had to express other non-unit fractions as combinations of unit fractions.

3 **Show that $\frac{5}{36} = \frac{1}{12} + \frac{1}{18}$**
4 **Find the missing number: $\frac{2}{7} = \frac{1}{\square} + \frac{1}{28}$**
5 **Investigate how to convert fractions into a combination of unit fractions, in the Egyptian way.**
   Some fractions may require more than two unit fractions.
   Can you find a general rule? Make sure to verify your rule, if you find one.

◆ Assessment opportunities

◆ In this activity you have practised skills that can be assessed using Criterion B: Investigating patterns.

How do we construct a number line and why?

▼ Links to: Biology, Citizenship and Inclusion

Naoki is autistic. What is autism? How does autism affect how we see the world? What does neurobiology tell us about brain organization and why some autistic students show special talents in mathematics?

DO YOU HAVE A FAVOURITE NUMBER?

Naoki Higashida, an author, links the discovery of numbers to proof that we humans are creative and insightful. He uses the discovery of zero as proof that humans are civilized. Do you agree with him?

Read 'What's your favourite number?' at **http://time.com/4856602/autism-nonverbal-book-naoki-higashida/** and decide whether your favourite number is:

- zero, which stands for nothing
- one, which indicates something
- two, which allows us to divide and sort things
- three, which, according to Naoki, was invented though it wasn't needed
- or another larger (or smaller) number.

Why is your favourite number significant to you?

THE CONCEPT OF ZERO

If we have quantities or numbers, then it follows naturally that some are bigger or smaller than others. We know that some cultures did not use numbers but simply employed the ideas of 'one' and 'many'. Others have a few numbers, but none beyond a certain point.

We saw in *Mathematics for the IB MYP 1*, Chapter 1, that numbers can be placed on a number line, with each whole number an equal distance apart. If we know that numbers continue counting up to the right, it is logical that they continue to the left as well, as shown in the diagram.

Each step to the left decreases the number by 1. Once humans were working with increasing numbers by counting, it is easy to see why they began to contemplate decreasing numbers. If 2 is one less than 3 and 1 is one less than 2, what is less than 1?

What is zero? If we have something and we can lose it, there must be a concept of nothingness. But is zero the same as nothing? Does it indicate nothing or is it a number with a value of zero? Does zero on an elevator panel mean there is no floor there? Does zero degrees Celsius (0°C) indicate a lack of a temperature?

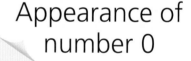
Appearance of number 0

Humans have wrestled with the idea of zero for a very long time. Both the Mayan people and the Indian mathematicians discovered and used zero. The Mayans used the image of an empty oyster shell as one of their symbols for zero, to show the emptiness inside the shell. Look at our own symbol for zero, 0. Is it representing nothing or does it show nothingness inside it?

We take zero for granted. Humans use it every day and we are so used to it that it is very difficult to imagine a time when the symbol did not exist. But we haven't always had it. In fact, initially there was a lot of confusion about the concept and how it worked. Clearly, now we know that adding zero or subtracting zero from a number leaves it unchanged.

So, $a + 0 = a$

And, $a - 0 = a$

Explain why this is so.

We also know that $a \times 0 = 0$

Is it possible to multiply a number by zero? Or divide? Explain your answers.

The great Indian mathematician Brahmagupta established the mathematical rules for dealing with zero, although he did incorrectly think that $1 \div 0 = 0$. We will read more about him later in this chapter.

THINK-PAIR-SHARE

So, how did those mathematicians who didn't have zero calculate? In some countries you hear 'nought' as well as zero to represent nothing. Where do you think these words came from?

The number zero was not common until the 12th century. It had a long and controversial history, as far as numbers go. To hear more about it, listen to The Curious Cases of Rutherford and Fry, www.bbc.co.uk/programmes/p04plxmj.

THINK-PUZZLE-EXPLORE

Consider the following questions and discuss them in groups:
- **What is the largest number you can think of?**
- **How long would it take you to count to that number, if you were counting two numbers per second?**
- **What is at the end of the number line?**
- **What is the concept of infinity? When was this idea first discussed?**

FROM NOTHING TO EVERYTHING

Does counting have a limit? If you keep counting upwards, you will eventually get bored because it is time-consuming and repetitive. Yet huge numbers do exist, like the number of leaves in a forest or people in a city.

Can we say that infinity was discovered or created? It definitely perplexed the Greek mathematicians! Leonhard Euler (1707–1783) discussed infinity as a number but never explained what he meant by that.

Georg Cantor (1845–1918) is remembered as one of the more modern mathematicians, to examine infinity. He laid out a number of theories and concepts about infinity. One of his fundamental concepts was that an infinite amount does not have to take up an infinite space; for example a single circle contains an infinite number of points and rotations.

The infinity symbol certainly **was** created. John Wallis (1616–1703) first used the infinity symbol ∞ (sometimes called the **lemniscate**). But what does the symbol actually have to do with infinity?

WHAT ABOUT OTHER 'BIG NUMBERS'?

Sometimes we use words to show large numbers such as million, billion and trillion. But we also have other, non-mathematical words such as gazillion, bajillion and umpteenth. These types of words have a name: **indefinite hyperbolic numerals**!

What purpose do these large, non-specific numbers have? We use them to suggest an unknown but huge number. To learn more about them, listen to *More or Less* at this link: **www.bbc.co.uk/programmes/b093hf8v**

Some people just don't like exaggeration or hyperbole. Statements like 'I've listened to that song a billion times already' make them want to check if that feat is even possible. Search for these sometimes silly calculations and look out for they did the math.

ACTIVITY: First estimate, then calculate

■ ATL

- Critical-thinking skills: Draw reasonable conclusions and generalizations

Estimate, then calculate, the answers to these questions:
- **Can you describe what a million means?**
- **Can you see a distance of 1 million millimetres?**
- **How heavy is one million dollars?**
- **How much space will be taken up by one million centimetre-cubed blocks?**
- **What length of time is one million seconds?**

◆ Assessment opportunities

- In this activity you have practised skills that are assessed using Criterion A: Knowing and understanding.

▼ Links to: Design

The lemniscate is very popular in logo design, although companies and brands like to change the colours and sizes of the loops. Why do you think they like to use the symbol or concept of infinity?

Infinity Records

Where the vinyl never ends

How do we operate with numbers?

ARITHMETIC AND OPERATIONS

We saw earlier in the example of Ali the trader, on page 8, that he needed addition and subtraction when he bought and sold at the marketplace.

The + and – signs were Latin terms meaning 'more' or 'less'. In a book called *Mercantile Arithmetic* the first appearance of a minus sign was seen in 1489. Robert Recorde (1512–1558) introduced the modern use of minus in 1557. Use addition to predict the correct answer, and then check along the paths.

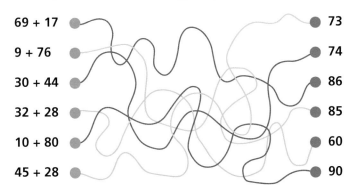

We use some special words to describe the numbers in sums.

- summand + summand **gives a** ● sum or total
- augend + addend + addend **also gives a** ● sum or total

(The augend is the first number in a string to be added.)

One of the first descriptions of an addition calculation comes from the 12th century by the ancient Indian mathematician called Bhaskara. In a book written to his daughter Lilivati, he sets out a problem for her to find the sum of 2, 5, 32, 103, 18, 10 and 100 added together. Adding units, tens, hundreds, and so on, quickly became the easiest way to find sums (or totals). Use subtraction to predict the correct answer, and then check along the paths.

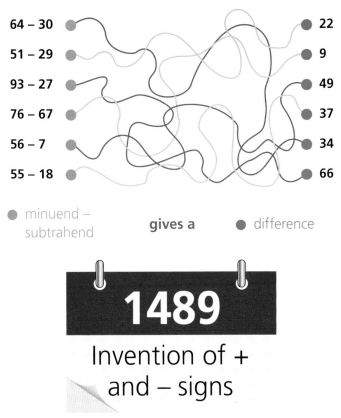

- minuend – subtrahend **gives a** ● difference

1489

Invention of + and – signs

Subtraction is a harder skill and was discovered later. Subtracting means finding the difference between two numbers, and this can be done by calculating in 'jumps' to get from the first number to the second.

For example, what is the difference between 58 and 762?

We can see that to get from 58 to 762 is 704. The more traditional way to show this is:

$$
\begin{array}{r}
7\,6\,2 \\
-\ \ 5\,8 \\
\hline
7\,0\,4
\end{array}
$$

Without using adding or subtracting signs, the Greeks operated with numbers on an abacus. There is evidence to suggest that the first mechanical calculating device was already being used in Babylon in 500BCE.

MULTIPLICATION

Use multiplication to predict the correct answer and check along the paths.

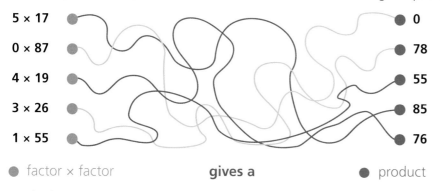

5 × 17   0
0 × 87   78
4 × 19   55
3 × 26   85
1 × 55   76

● factor × factor   **gives a**   ● product

'Multiplication is nothing more than repeated addition.' What is meant by this statement?

If 7 × 12 is '7 lots of 12', then the answer could also be found by 12 + 12 + 12 + 12 + 12 + 12 + 12.

However, this is a time-consuming way to find the answer, so multiplication was developed.

In *Mathematics for the IB MYP 1*, Chapter 6, we looked at a method of multiplication used by the Egyptians, which is called Duplation. This involves doubling and doubling to find combinations of multiples that can be combined to give a product.

For example:

$$73 \quad \times \quad 65$$

1	65
2	130
4	260
8	520
16	1040
64	4160

$$73 = 64 + 8 + 1$$

$$\therefore 73 \times 65 = 4160$$
$$_{1}520$$
$$+ \quad 65$$
$$\overline{4745}$$

In the late 17th century, there were at least 10 different ways to represent multiplication. Today multiplication is commonly shown by one of three signs: ·   ×   *

An **interpunct** is the dot (·) that is used to represent multiplication to avoid confusion between × (multiplication sign) and *x* (the unknown or variable in algebra).

1631
Invention of the × sign

POWERS OR EXPONENTS – MULTIPLYING BY SELF

If multiplication is a 'trick' or shortcut for addition, then powers (or exponents or indices) are a shortcut for multiplication when repeated multiplying by self is involved.

$3 \times 3 = 9$

$3 \times 3 \times 3 =$ $\quad\quad 9 \times 3 = 27$

$3 \times 3 \times 3 \times 3 =$ $\quad 9 \times 3 \times 3 =$ $\quad\quad 27 \times 3 = 81$

These sums are fairly manageable to write out, but what if you had to calculate **3** tripled **15** times?!

The short version to show this is

3^{15}

Base   The number of multiplications of the number by itself

DIVISION (PARTITIONING OR SHARING)

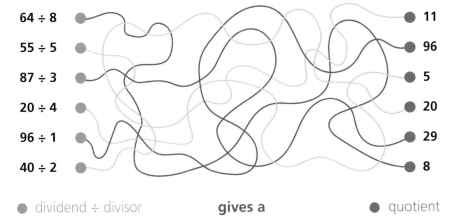

$64 \div 8$		11
$55 \div 5$		96
$87 \div 3$		5
$20 \div 4$		20
$96 \div 1$		29
$40 \div 2$		8

● dividend ÷ divisor   **gives a**   ● quotient

While multiplication, subtraction and addition were being commonly used by merchants and clerks throughout the Middle Ages, division came along much later.

PRACTICE EXERCISE

1 Find:

  a 9^2

  b 6^3

  c 1^{101}

2 Use a calculator to find:

  a $4 \times 4 \times 4 \times 4$

  b 20^2

  c 5^4

3 Write out the following multiplications using exponents instead of multiplication:

  a $13 \times 13 \times 13 \times 13 \times 13$

  b $12 \times 12 \times 12 \times 12 \times 12 \times 12 \times 12 \times 12 \times 12 \times 12 \times 12 \times 12 \times 12 \times 12 \times 12$

  c $-2 \times -2 \times -2 \times -2 \times -2$

4 Find the values for each of the questions above.

EXTENSION

How would you handle an exponent in an exponent? Consider this example:

3^{2^3}

The division sign that we most commonly use is known as the **obelus** (÷). It is different from the other arithmetic operators in that it is a visual reminder of what is happening, where the dots above and below the line look like numbers in a fraction (or two numbers to be divided). We have explored a variety of division methods in previous books in the *Mathematics for the IB MYP* series, and almost all of them rely heavily on knowledge of multiplication and times tables.

WHAT'S LESS THAN NOTHING?

The Nine Chapters on the Mathematical Art (Jiǔzhāng Suànshù) is a Chinese mathematics book, composed by several generations of scholars over centuries from the 10th to the 2nd century BCE. This ancient book is one of the oldest surviving mathematical texts and shows an approach to mathematics that centres on finding the most general methods of solving. Negative numbers are addressed in this book, although they would be ignored by Western scholars for centuries after this.

Mathematicians in the Middle Ages thought equations such $x + 3 = 0$ couldn't be solved, and the negatives they could see in (ancient) Islamic texts had no use or meaning. How would you convince them that they were wrong?

The mathematician Brahmagupta records negative numbers in India in the 7th century, thanks to debts accrued by banking, and he also refers to them as solutions to quadratic equations – that is, equations where the highest power of the unknown is 2.

ABSOLUTE VALUES

The idea of absolute value describes the distance of a number from zero without mentioning if it is that distance above or below zero – so it is the value of the number irrespective of whether it has a positive or negative value.

The absolute value of 8 is 8; the same is true for the absolute value of −8.

The notation for absolute value places the number between two vertical lines:

$| \; |$

For example:

$|-2| = 2$	$|-2 + 3| = 1$
$|2| = 2$	$|-2 - 5| = 7$

PRACTICE EXERCISE

Solve:

1 $|-3| =$

2 $|-3 + 8| =$

3 $|11 - 3| =$

4 $|-12^2| =$

5 $|-12^2 - (-4 \times 3)| =$

6 $-2|9^2| =$

7 If $|x^2| = 100$, what are two possible values for x?

8 Plot all the answers to the questions above on a single number line. Watching this video may help you: www.khanacademy.org/math/algebra-basics/basic-alg-foundations/alg-basics-absolute-value-new/v/absolute-value-and-number-lines.

Why do we find rules and then try to justify them?

There comes a point in many cultures where people stop using mathematics for practical purposes and begin to play with it, for its own sake or for fun. Another key development in the history of mathematics occurs when relationships are turned into general rules and formulae, which are then tested and justified. This is a huge leap from utility (usefulness) into abstraction (moving from specific events or examples into general rules).

There were several important milestones on the road to algebra. In 1247CE in China, Qin Jiushao (1208–1261) wrote *Shùshū Jiǔzhāng* (*Mathematical Treatise in Nine Sections,* not to be confused with *The Nine Chapters on the Mathematical Art*) in which he included equations and the numerical solution of certain types of equations.

In Europe, the leap of moving from knowns to unknowns coincided with the Renaissance. Girolamo Cardano (1501–1576) is well known for his contributions to the branch of algebra. He made the first systematic use of negative numbers, and the stories of his disagreement with another mathematician, Tartaglia (1449–1557), make for very interesting reading.

1557

Invention of = sign

THE WHETSTONE OF WITTE

In this influential mathematical book, *The Whetstone of Witte*, Robert Recorde introduces us to the equals sign for the first time. You can see an example on the image to the right. How is it different to the type of algebra problem we see nowadays? How is it similar?

Can you 'decode' the mathematical communication in the image?

In another giant step for algebra, François Viète (1540–1603) introduced the use of letters to represent variables and constants. Find out more about François Viète and his fascinating life as a mathematician and codebreaker to a king.

1591

Introduction of letters in mathematics

ACTIVITY: Who invented algebra?

ATL

■ Critical-thinking skills: Consider ideas from multiple perspectives

The question 'who invented algebra?' was posted in early 2013 on an answers forum called quora.com: **www.quora.com/Who-invented-algebra**

People can submit answers and other readers can 'up vote' or 'down vote' them, depending on whether they find the answers useful or correct.

On the right, you can see a post from Professor David Joyce.

A student, Dimitra Varoufakis, reads Professor Joyce's post and disagrees. She decides to post on the discussion herself and argues that because the Greek mathematician Diophantus wrote a series of books on the subject known as Arithmetica, the invention of algebra is Greek. This took place around 200ce. She gets over 3000 views and 14 upvotes.

Another student, Somreeta Patel, thinks that algebra is even older than Diophantus and could be credited to al-Khwarizmi. However, she also knows there is more to algebra than the use of symbols, so the great Indian mathematicians Brahmagupta and Bhaskara, who developed procedures for solving equations without using symbols, could be considered the inventors. She wants to post this answer too. What would you recommend?

Read again and carefully consider David's, Dimitra's and Somdatta's arguments. Now it's time to decide … Who do you think invented algebra? Why?

Make a convincing argument to say who invented algebra. You will need to do additional research and reflect on the topic before you make your mind up. Use these search terms: history of algebra, who invented algebra.

Do you think someone who is Indian, Chinese or Arab might answer differently from someone who is not?

How might your nationality or identity affect your decision? Do you think pride (how proud or nationalistic you are) might play a part?

Can our identity affect our mathematical understanding or beliefs? How accurate would our decisions be then?

 David Joyce
Professor of Mathematics and Computer Science
Written Mar 13, 2014

It all depends what you mean by algebra.

Solutions to linear and quadratic problems

If by algebra you mean solving for unknowns when you know something about them, such as the thing squared plus three times the thing equals 144, that's ancient. The Babylonians knew how to solve linear and quadratic problems 4000 years ago. The Egyptians could solve many of those about the same time. Euclid encoded their solutions in the *Elements* 2300 years ago. The Chinese could solve them 2000 years ago or earlier. They could also solve systems of linear equations back then, too. So could the Indians then or earlier.

Methods to solve linear and quadratic problems were known in all the advanced ancient civilizations.

The word algebra comes from a word used by Muhammad ibn Mūsā al-Khwārizmi (ca. 750– ca. 850) in a book he wrote that systematically described the solutions of linear and quadratic problems by reducing them to simpler problems.

Symbolic algebra

If by algebra you mean symbolic algebra that has signs for the arithmetic operations and for equality, and symbols for variables, that was created in the 1500s and continued to be standardized in the 1600s. No one person invented that. You can point to individuals who contributed to the development of symbolic algebra, and a couple who were influential in standardizing and spreading its use, but you can't say it was the invention of any one person.

6.8k Views • View Upvotes

Upvote | 7   Downvote Comment

◆ Assessment opportunities

◆ In this activity you have practised skills that are assessed using Criterion C: Communicating.

How do our surroundings affect our identities as mathematicians, and vice versa?

As we discovered in our investigation of Newgrange, ancient people understood something about the mathematical relationships of the Sun and the time of year, predicting the angle of sunlight at the solstice, which is a reoccurring phenomenon. Discovery in geometry often occurs when we use mathematics to learn more about the space we occupy, and to express and reflect on what we already know.

By 1900BCE, Egyptians were using the beginnings of geometry to apply to problems, which were often inspired by their environments. The needs of navigation, religion, astronomy, and surveying and building drove mathematical developments.

One such example occurred in ancient Egypt where a relationship between the star, Sirius, and the regular flooding of the Nile shortly after its appearance was noted.

By recording the appearance of Sirius over the horizon just before sunrise on a summer morning for many years, Egyptians could predict the flooding of the Nile. They came very close to working out the length of a year that we use now.

It seems almost incredible that such complex calculations were being done so long ago, before people knew about the signs and operations we use today. Here are some incredible discoveries in astronomy:

- **4236BCE**   The calculation of the length of a year
- **428BCE**   The calculation of the distance between the Earth and the Moon
- **200BCE**   The calculation of the circumference of the Earth
- **1851CE**   Discovery of the rotation of the Earth

ACTIVITY: Calendar crazy

■ ATL

- Information literacy skills: Evaluate and select information sources and digital tools based on their appropriateness to specific tasks

Many peoples of the ancient world used calculations and observation to create calendars. They used them to record the movement of their gods (celestial bodies), to anticipate cycles, to plant crops, celebrate festivals and to conduct the regular cyclic affairs of the kingdom. This is exactly what we use these calculations (calendars) for, even today.

Research the calendars of the following civilizations. How did they work and how accurate were they?

- **The Yoruba farming cycles**
- **The Mayans**
- **The Khmer temple**
- **The Calendar of Troth**

◆ Assessment opportunities

- ◆ In this activity you have practised skills that are assessed using Criterion D: Applying mathematics in real-life contexts.

How are measurements 'born' and why do they die?

Measurement is fundamental to construction on large and small scales. If we measure something, the value is meaningless without a unit. Units and measurements were essential to the Egyptians, especially with regards to their incredible ancient pyramids.

How tall are you? How long is your hair today? How much have you eaten this week? Answer all these questions as accurately as you can. What units did you use? How accurate are your answers?

You may remember from *Mathematics for the IB MYP 1*, Chapter 5 that long before money was invented, ancient peoples would exchange goods and services through bartering (trade). The people of the Indus Valley – today's Pakistan, West India and north eastern Afghanistan – adopted a standardized form of **mass measurement** by comparing various weights. Babylonians used their arms and fingers to describe lengths, and empty containers were filled with plant seeds which were then counted to describe volume. Egyptians and Mesopotamians eventually created the first unit of measurement, the 'cubit', using a cubit rod whose length extended from the elbow to the tip of the middle finger. This was not entirely consistent, as arm length varies, but gave a good estimate. Denominations of the cubit included palms, fingers, and even finger nails. This led to the eventual creation of inches, feet and yards, and the Romans later added the mile (*mille passus* meaning 1000 paces). You may remember this from Chapter 1 of *Sciences for the IB MYP 2: by Concept.*

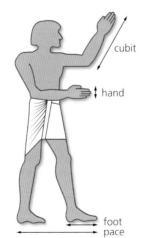

cubit

hand

foot
pace

By the 19th century, the most accurate units of measurement were found in what is now known as the **Imperial** (coming from the word *empire*) **System**. This system was widely used across the countries that had been colonized by the British Empire. A growing Empire required a great deal of agreement, for example in mapping and navigation. Therefore, a way to compare quantities, no matter the country or context, was needed.

In the USA, Myanmar and Liberia these units are still used today. Some Imperial units such as pints, feet and inches are also still used in the UK and some Commonwealth countries. The rest of the world uses the metric system, which is a system of increasing units in the orders of magnitude of (or powers of) 10. It is possible to convert between Imperial and metric units using formulae or conversion charts.

ⓘ **1 m** = 3.28084 feet

PRACTICE EXERCISE

Convert the following quantities (questions 1–8).

1   10 metres into feet

2   250 metres into feet

3   4.7 metres into feet

4   12 feet into metres

5   5.5 feet into metres

6   921 cm into feet

7   17.5 feet into metres

8   25 000 feet into kilometres

9   If Ali is 1.76 m tall, Elisaveta is 6 foot 3 inches tall, Aleksander is 4 cubits tall, who is the shortest and who is the tallest?

MEET A MATHEMATICIAN: PLATO (428/7BCE–348/7BCE)

Learner Profile: Communicator

Plato is possibly one of the most famous Greek philosophers of all time. While he was hugely influential as a philosopher and teacher, he was also a mathematician who is said to have had the famous quotation shown in the picture (left) written at the entrance to his school. He was associated with the Pythagoreans, who you will meet in Chapter 5. One of his greatest contributions to mathematics is his identification of the **Platonic solids**, which he described around 380BCE.

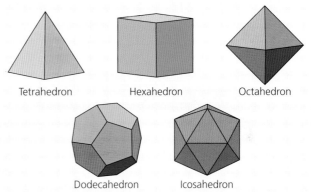

| Tetrahedron | Hexahedron | Octahedron |

Dodecahedron   Icosahedron

These solids are three-dimensional shapes whereby each face has the same regular polygon and the same number of those polygons meet at each vertex or corner. All five Platonic solids are shown and named above. Verify that they meet the description (or conditions) of a Platonic solid.

We will study some of these concepts of geometry and trigonometry in Chapter 5, where we make use of these ideas to find lengths, angles and areas in space.

BUT WHY IS AGREEMENT IMPORTANT?

Although metric units were invented in 1795, the use of Imperial units was widespread and dominant throughout the 19th century. As scientific inquiry became more collaborative as well as competitive, the need for standard units became more and more important. If an experiment cannot be repeated or recreated, it doesn't hold up to the standard of the scientific method.

The need for a unified system that could be used internationally led to the International System of Units, or **SI Units** for short. Adopted by the 11th General Conference on Weights and Measures in 1960, the following units were exactly defined and accepted as standard:

- ampere
- kelvin
- mole
- second
- candela
- kilogram
- metre.

You will already be familiar with some of these units, and you will use others in your learning in Sciences and Mathematics through the Diploma Programme.

Find out what each unit is used to measure and the exact definition of a single unit for that quantity.

So now we know measurements are only useful if they are commonly used and their value is agreed. But what about measurements that were used and agreed in the past, but not anymore?

- A cawnie of land
- A koku of rice
- An uncia of silk

ACTIVITY: Tempus fugit

For this activity, you will be completing research and considering the nature of time itself. This is a really big idea. Now we take for granted that time (an hour, a minute, a second, the time shown on a clock) is something that is regular, agreed and reliable. But this wasn't always the case. Once humans didn't use markers of time, as we know them. Not all that long ago, the time on a town clock in one village was probably different from that of neighbouring villages, not even that far away.

Visit this list of obsolete (no longer used) units to learn more about them: **https://en.wikipedia.org/wiki/List_of_obsolete_units_of_measurement**

Here are some very important moments in our mathematical development of measurement.

4236BCE	The calculation of the length of a year
428BCE	The calculation of the distance between the Earth and the Moon
200BCE	The calculation of the circumference of the Earth
1862CE	The calculation of the speed of light
1953CE	Discovery of the structure of DNA

Were each of these discovered or invented? Who was responsible for each development? How and why? Were these people's motivations the same or different? Were their methods similar or different? Are any of them 'debatable discoveries' (could you argue this is **not** when they were discovered or invented)?

Your task is to carry out research to answer the following questions:

- **Before clocks, what were the different methods of timekeeping? Give examples.**
- **Do we still use any of those methods today?**
- **What did the invention of trains have to do with measuring time?**
- **How and when were time zones created?**
- **What is an atomic clock and what is it used for?**

You may choose to represent your findings as an article, a presentation, a poster or a comic strip. Be sure to use mathematical language and notation, where appropriate.

Measurement also led us to inquire into change, and more importantly rates of change. The study of these rates changes opened up the field of **calculus**, which you will learn about in future studies (Diploma Programme Mathematics).

$$\text{gradient} = \frac{dx}{dy}$$
$$\text{area} = \int f(x)dx$$

PRACTICE EXERCISE

Use the globe photo and your research skills to answer these questions. You can examine the photo in more detail at this link **https://bit.ly/2k0VRfM**

1   If it is midnight in New York, what time is it in Moscow?

2   If it is 11:30 a.m. in Addis Ababa, what time is it in Kuala Lumpur?

3   What are we adding to? What does the plus sign mean?

4   If it is 15:22 in Lisbon, Portugal, what time is it in Marbella, Spain?

5   At a quarter to eight in the morning in Melbourne, Australia, what time is it in Jaipur?

6   How many countries have more than two time zones?

7   In what region is the +5:45, as seen on the globe?

8   Some of the time differences change throughout the year. Why is this? Is it consistent?

9   If it was agreed suddenly that Afghanistan became the central time zone (+0) and all other time zones had to change, what would the new globe markings become?

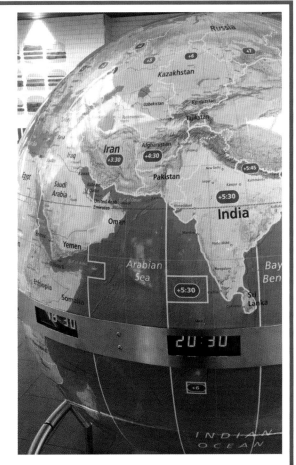

■ This photograph shows a globe display in Schiphol Airport, Amsterdam. What mathematics can you see here?

How can we use mathematics to reveal past or predict future behaviours?

MATHEMATICS FOR FUN AND LEISURE

People have always enjoyed riddles, puzzles and games, and many of these are mathematical in nature. Very few people, outside academic circles, had the time or wealth to indulge in deep mathematical thinking until the 1700s. At that time, leisure (free time) was a relatively new concept and the number of people playing with mathematics suddenly increased.

During the time of empires and into the Industrial Age, there was an increase in the number of men (and sometimes women) who engaged in recreational mathematics, whether for amusement, innovation or education.

Free time and wealth did not always lead to academic interests, of course. Many wealthy young men (and sometimes women) turned to games and gambling for entertainment. This coincided with, maybe even led to, the birth of the **probability** branch of mathematics. Players and gamblers wanted ways to understand possible outcomes, to judge if a game was fair or not, and to try to predict the future.

VENN DIAGRAMS AND SETS

The 19th century was also an exciting time in the invention of different types of data representation, and in the 20th century these visuals became regularly used in newspapers and in business.

In mathematics, a group or collection of numbers or objects is known as a **set**. One method of representing different sets or groups was created by John Venn (1834–1923) around 1880, although he called the diagrams **Eulerian circles**. We used these to represent factors and multiples in *Mathematics for the IB MYP 1*, Chapter 6.

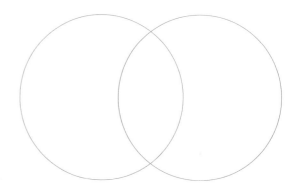

By creating overlapping circles, the objects or numbers can be placed in the diagram to show which members of the set share characteristics and which do not. Let's begin with two different sets:

- Set A – the even numbers between 0 and 11 (2, 4, 6, 8, 10)
- Set B – the factors of 100 (1, 2, 4, 5, 10, 20, 25, 50, 100).

These numbers can now be placed on a Venn diagram like this:

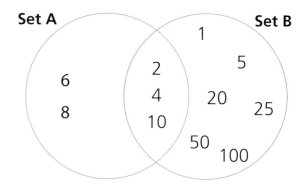

Note that 2, 4 and 10 appear in the overlapping section because these numbers appear on both lists or in both sets.

If we have three sets, we can add another circle to represent the additional set:

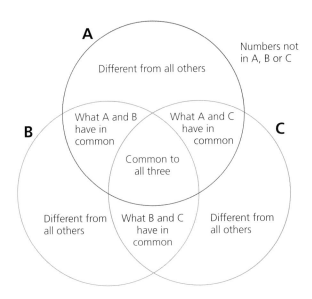

In the example (below left), we can see that the numbers 2 and 4 appeared in all three sets so they are marked in the central location. When two or more sets overlap, this is called the intersection. The symbol is ∩ and we indicate which overlaps or intersections we are looking for like this:

$$A \cap B \cap C = \{2, 4\}$$

To list **all** the elements in two or more sets we use a union symbol ∪. For the example:

$$A \cup B \cup C = \{1, 2, 3, 4, 5, 6, 7, 8, 10\}$$

Example

Place the following sets on a Venn diagram:
- Set A = (1, 2, 3, 4, 5)
- Set B = (2, 4, 6, 8, 10)
- Set C = (2, 3, 4, 5, 7)

Solution

Be careful to identify the members which appear in one set only, appear in two sets or appear in all three sets, and place them in the right places:

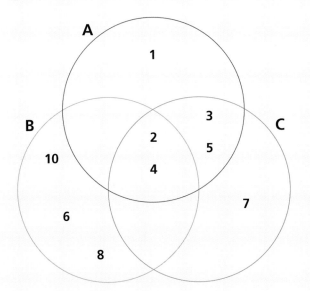

PRACTICE EXERCISE

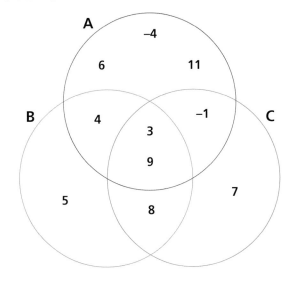

1   List all the elements of set A.

2   List all the elements of set B.

3   List all the elements of set C.

4   State A ∩ B ∩ C.

5   State A ∪ B ∪ C.

6   State A ∪ B.

7   State B ∩ C.

8   Find the set which only contains positive integers.

Venn diagrams are not only used for numerical examples. They can display objects as well as numbers. Here are some fun Venn diagrams that are used to illustrate information in a light-hearted way.

Pancakes

Flour · Pasta · Egg

Batter · Omelette

Milk

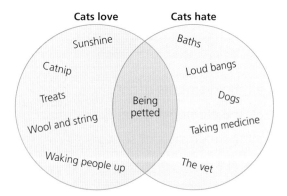

Venn diagrams can also display groups which are **subsets** of, or totally contained within, other sets. These types of diagrams can make complex identities and relationships much easier to see.

Venn diagrams are extremely useful for calculating probabilities of certain outcomes. For example, if 18 class members have volunteering, cello and ice hockey as afterschool activities, then we can identify how many people do which combinations of activities and we can calculate the chance of randomly picking someone from each group.

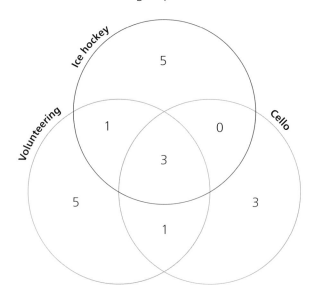

What is the probability (P) of a randomly chosen student participating in ice hockey? Find the number of students out of a total of 18 who play ice hockey.

5 + 1 + 3 + 0 = 9 students out of 18

P(ice hockey) = 0.5 or $\frac{1}{2}$

What is the probability of randomly choosing a student who both volunteers **and** plays cello?

volunteers ∩ cello = 4 students out of 18

P(volunteers ∩ cello) = 0.22 or 22% chance of choosing one of them

What is the probability of choosing a student who volunteers **or** plays ice hockey (or both)?

volunteers ∪ ice hockey = 15 students altogether out of 18

P(volunteers ∪ ice hockey) = 0.833 or 83%

Searching for truth in the past and in the future gave rise to the mathematical branches of statistics (past) and probability (future). These two areas developed hand in hand – as increasing amounts of data were gathered through statistics, there were more numbers to play with and more probabilities to find.

In 1654, Blaise Pascal (1623–1662) discovered the theory of probability by collaborating with Pierre de Fermat (1607–1665) over a gambling dispute. The discovery was prompted by a friend, the Chevalier de Méré, and a specific problem in which two players wanted to finish a game before its conclusion and wished to share the stakes (or money) fairly, given the chances each one had of winning the game if they had played on from that point.

Here is a mathematical definition of probability: 'The extent to which an event is likely to occur, measured by the ratio of the favourable cases to the whole number of cases possible.' We will see a lot more about probability in Chapter 4.

Take action

! Get involved in citizen mathematics. There are lots of online opportunities to contribute to the discovery of new primes. Programmes such as GIMPS at **www. mersenne.org/** and PrimeGrid **www.primegrid.com/** are called **distributive computing projects**. These allow you to dedicate unused processing time to calculation, so computers can work together to find the next prime number. Prime numbers are interesting mathematically but also currently very useful in online security and digital photography.

1654

Invention of probability theory

What's new in mathematics?

In 1982, Benoît Mandelbrot published an exciting new development in geometry explaining the concept of fractals, with applications to diverse natural phenomena. If we think of a coastline on a map, we can picture variation along the coast. If we zoom in, this variation doesn't smooth out to give us less variation. Instead, zooming in reveals more detail and more variation. This keeps happening the more you zoom in or enlarge the coastline.

WHAT IS A FRACTAL?

A **fractal** is defined as 'a curve or geometrical figure, each part of which has the same statistical character as the whole'. They are geometrically repeating structures, so that if you zoom in, similar patterns are revealed at smaller and smaller scales. These are modelled and drawn exceptionally well by computers and have contributed to the improvement in computer animation as fractals are used to create many nature-like effects. The Disney movie *Frozen* even refers to them in one of the hit songs: 'Spiralling in frozen fractals all around'. You can find many hypnotizing images of the Mandelbrot set fractals.

Fractals are useful in modelling structures (such as snowflakes) in which similar patterns recur at progressively smaller scales and in describing partly random or chaotic phenomena, such as crystal growth and galaxy formation.

1982
Geometry of fractals published

MEET A MATHEMATICIAN: MATT PARKER, BORN 22 DECEMBER 1980

Learner Profile: Communicator

Matt Parker is an Australian mathematics communicator, stand-up comedian, writer and teacher. Let's hear about Matt from his own website:

> 'Possibly the only person to hold the prestigious title of London Mathematical Society Popular Lecturer while simultaneously having a sold-out comedy show at the Edinburgh Festival Fringe, Matt is always keen to mix his two passions of mathematics and stand-up.

> Originally a maths teacher from Australia, Matt now lives in the UK and works both as a stand-up comedian and a maths communicator. This involves spreading his love of maths via books, YouTube videos, radio programmes, TV shows, school visits and live comedy shows. Matt is also the Public Engagement in Mathematics Fellow at Queen Mary University of London.'

Source: http://standupmaths.com

You can follow Matt on Twitter **@standupmaths**, or watch him on YouTube at this link: **youtube.com/ standupmaths**.

Recently Matt also wrote a book of recreational mathematics called *Things to Make and Do in the Fourth Dimension*. And if you get the chance to meet him, Matt will dedicate a copy of his book (sign with your name in) but he will instantly transform your name into **binary** instead of letters! For more about binary systems see *Mathematics for the IB MYP 2*, Chapter 6, or later in this chapter. Matt is also passionate about **Woodall primes**. More about these, and prime numbers in general, can be found in *Mathematics for the IB MYP 3: by Concept* Teaching and Learning Resources.

SUMMATIVE ASSESSMENT

Use these problems to apply and extend your learning in this chapter. The problems are designed so that you can evaluate your learning at different levels of achievement in Criterion C: Communicating.

The Statement of Inquiry for this chapter is:

Firstly humans observed phenomena and relationships. Then they measured quantities. Soon they could create general rules and formulae which could be justified. All these ways of knowing come together to give us our mathematical body of knowledge.

Throughout the whole chapter we have been looking at discoveries and contributions made by cultures and individuals over time. Those selected are only a sample of all the ways of knowing that have developed through time. Your task is to use the information provided, and additional research, to communicate an overview of the history of mathematics.

You might wish to consider using:

- a timeline showing the major discoveries and crediting them to societies or individuals. This timeline could be static, like a poster or display, or it could be technology-enabled using timeline software such as www.tiki-toki.com/
- a map showing contributions coming from different cultures and countries, using a globe or graphic like the one in the picture above
- an augmented reality tour, in which people can access facts and figures by reading posters or moving along a timeline display. You will need to be tech-savvy for this option!

Reflection

Questions we asked	Answers we found	Any further questions now?
Factual: How do we construct a number line and why? How do we operate with numbers? What's new in mathematics?		
Conceptual: How did numbers take form and shape? Why do we find rules and then try to justify them? How are measurements 'born' and why do they die?		
Debatable: What came after counting? How do our surroundings affect our identities as mathematicians, and vice versa? How can we use mathematics to reveal past or predict future behaviours? Is mathematics discovered or invented? Is a more commonly used unit the best one to go with?		

Approaches to Learning you used in this chapter:	Description – what new skills did you learn?	How well did you master the skills?			
		Novice	Learner	Practitioner	Expert
Communication skills					
Critical-thinking skills					
Information literacy skills					
Creative-thinking skills					
Learner Profile attribute(s)	Reflect on the importance of being a good communicator for your learning in this chapter.				
Communicator					

2 How do we make choices?

Real-life problems can be **represented** by different **forms** of mathematics which will yield **equal** results and a **fair** solution.

○ IN THIS CHAPTER, WE WILL ...

- **Find out** about linear relations.
- **Explore** how to model relationships using algebra, tables of values, and graphs.
- **Take action** by developing linear equations and graphs to identify the most efficient fundraising activity, then carry it out to complete a very specific goal for a cause of your choice.

CONSIDER THESE QUESTIONS:

Factual: How do I know it's linear? How do I determine a rate of change? What is a fixed cost?

Conceptual: Can different expressions be equal? What are the different ways of solving an equation? Can a constant be graphed? How can linear equations influence our choices?

Debatable: Can all problems be represented by graphs? When is standard form a better alternative?

Now **share and compare** your thoughts and ideas with your partner, or with the whole class.

■ These Approaches to Learning (ATL) skills will be useful ...

- Organization skills
- Creative-thinking skills
- Critical-thinking skills
- Transfer skills
- Communication skills
- Information literacy skills

◆ Assessment opportunities in this chapter:

- **Criterion A**: Knowing and understanding
- **Criterion B**: Investigating patterns
- **Criterion D**: Applying mathematics in real-life contexts

● We will reflect on these Learner Profile attributes ...

- **Knowledgeable** – We develop and use conceptual understanding, exploring knowledge across a range of disciplines. We engage with issues and ideas that have local and global significance.

- **Reflective** – We thoughtfully consider the world and our own ideas and experience. We work to understand our strengths and weaknesses in order to support our learning and personal development.

Reflect on what you already know about:

- how to simplify algebraic expressions and solve two-step algebraic equations
- how to plot ordered pairs on a Cartesian plane from a table of values.

veterans	cost
profit	earnings
loss	base wage

SEE-THINK-WONDER

Take a close look at this image.

What do you see in this image? What do you think about it? What does it make you wonder?

What are the different ways of solving an equation?

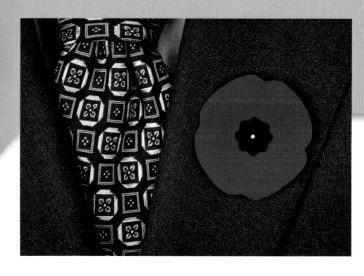

Molly is selling poppy pins to raise money for Armed Forces veterans and to honour fallen soldiers. She sells them for 25 cents each and hopes to raise $300 by the end of the week. How many poppies should she order to ensure she has enough to reach her goal?

SUBSTITUTION

Let's start by defining a variable. The question above indicates that we are looking for the number of poppies needed to raise $300. Thus, let p represent the number of poppies needed to raise $300. (Why would it be poor communication to state 'let p represent poppies'?)

We know each poppy is 25 cents (or $0.25), so the total money raised – the number of poppies times their price – is:

$0.25p$

We can substitute various numbers into p until we reach 300 – a process known as **trial and error**. Let's organize our efforts into a table of values.

p	$0.25p$	Result
500	0.25 (500) = 125	too low
1000	0.25 (1000) = 250	still too low
1500	0.25 (1500) = 375	too high
1250	0.25 (1250) = 312.5	getting close …
1200	0.25 (1200) = 300	**bingo!**

We got there, but it took a few trials and could take even longer with more complicated numbers. Surely there is a faster way?!

SOLVING AN EQUATION

$0.25p$ is simply an expression that tells us how to calculate Molly's profit. Since we know the goal is $300, we can turn this expression into an equation:

$0.25p = 300$

Now we need simply to remember the algebraic rules of inversion to solve the problem:

$$0.25p = 300$$
$$\frac{0.25p}{0.25} = \frac{300}{0.25}$$
$$p = 1200$$

DISCUSS

Why, after dividing, did we get a bigger number than we started with?

Example: Two-step equation

The value of Sanjay's $21 000 car goes down by $2000 every year. Use an equation to determine when his car will become worthless (that is, when it has a value of $0).

If we let y represent the number of years since Sanjay purchased his car, then:

$21\,000 - 2000y = 0$

1   What does y represent in this equation?

2   Why is the expression equal to zero?

Example: Expanding

In some restaurants, servers must add all their tips together and divide them equally. If Cora earns ¥80 for her shift and gets 10% of the collective tips (total tips among all 10 servers) – let's call that t – she earns:

$$80 + 0.1t$$

1 Why is 0.1 the coefficient of t?

2 If Cora earns a total of ¥175, how much did all of the tips add up to that night? Justify your response algebraically.

3 What is the collective tip if the restaurant has to pay out a total of ¥1800 in tips in a single evening?

Solution

1 Calculating the percentage of a number means changing the percentage into a decimal and multiplying. 10% as a decimal is 0.1, and if we multiply that by t we will get the amount of money Cora gets from the collective tips.

2 If she earns ¥175, then:

$$80 + 0.1t = 175$$
$$0.1t = 175 - 80$$
$$\frac{0.1t}{0.1} = \frac{95}{0.1}$$
$$t = 950$$

Thus, the collective tip for the evening is ¥950. This means when all the servers put their tips together, they add up to ¥950. Each server received 10% of this amount.

3 If each server earns $80 + 0.1t$, then 10 servers would earn:

$$10(80 + 0.1t)$$

And since the restaurant has paid a total of ¥1800,

$$10(80 + 0.1t) = 1800$$

Expanding, we get:

$$10(80 + 0.1t) = 1800$$
$$800 + t = 1800$$
$$t = 1800 - 800$$
$$t = 1000$$

Thus, there is ¥1000 in the collective tip jar that night.

DISCUSS

Why do some restaurants choose to divide the tips equally among the servers? Is it fair? Consider a situation in which this may not be fair.

PRACTICE EXERCISE

1 $-3x = 1800$

2 $g - 400 = -100$

3 $13a + 7 = 39.5$

4 $-p + 2 = 10$

5 $14(5f + 19) = 32$

6 $-(8 - c) = -8$

3 Solve for y.

4 Why wouldn't a negative value for y make any sense?

Solution

1 The letter y represents the number of years since the car was bought. If we multiply 2000 by the number of years, we will see how much the car has gone down in value.

2 Since the expression to the left of the equation represents the value of the car in a given year, and since Sanjay wants to know when that value will be zero, we write a zero on the right side of the equation.

3 $21\,000 - 2000y = 0$

$$-2000y = 0 - 21\,000$$
$$\frac{-2000y}{-2000} = \frac{-21\,000}{-2000}$$
$$y = 10.5$$

4 A negative y-value would imply negative time. This could be the time before Sanjay bought his car, but we specified that y represents the number of years **since** he bought his car, so time begins at zero and increases.

Can different expressions be equal?

Now that we have reviewed substitution into expressions and solving equations, we are ready to combine these two ideas. Let's examine a scenario that involves **two** expressions at the same time.

Audrey works at a grocery store and Edward babysits. How long will they each have to work to earn the same amount of money?

First you need to ask: How much do they each earn per hour?

Audrey earns 12 euro per hour and charges a base rate of 9 euro to cover a meal and transport, and Edward earns 15 euro per hour.

Audrey's earnings

Let h represent the number of hours Audrey works:

$12h + 9$

Edward's earnings

Let h represent the number of hours Edward works

$15h$

Since we want to know when their earnings will be equal, we can write:

Audrey's earnings = Edward's earnings

$12h + 9 = 15h$

Wait! There are variables on **both** sides.

We solve this like any other equation: collect like terms and isolate a variable. This time we will collect like terms by collecting all terms with a variable on one side of the equals sign, and all constants on the other. Since the right side has no constants, it is simplest to move the $12h$ to the right. We do this by subtracting $12h$ from both sides:

$$\begin{array}{c} -12h \quad -12h \\ 12h + 9 = 15h \\ 9 = 3h \\ 3 = h \quad \text{or} \quad h = 3 \end{array}$$

Thus, at 3 hours they earn the same amount of money.

Before going any further, let's practise a few more equations that have variables on either side of the equals sign.

PRACTICE EXERCISE

Solve the following equations.

One term on each side:

1 $2a = a - 1$

2 $6a = 5a + 2$

Two terms on each side:

3 $3b + 1 = 5b + 3$

4 $5 - p = 3 + 3p$

Three terms on each side with negatives and fractions:

5 $\frac{1}{4}k + 2 = \frac{1}{2}k - 3$

6 $2(x + 1) = 5(x - 5)$

ACTIVITY: Algebra laps

To play good football, you must be tactical and creative, and be able to think fast in a pressured situation. In the end, however, you still won't do very well unless you have the stamina to run for long periods – the average (mean) footballer runs over 11 km per game. Every professional footballer practises set pieces and trick shots, but they all also run laps in training. Here are some algebra laps to keep you in shape for what is to come.

Solve the following equations.

1. $5.1a = 12$
2. $c - \frac{1}{5} = \frac{1}{3}$
3. $\frac{k}{6} = 100$
4. $4x + 7 = 23$
5. $3.1t + 17 - t + 1 = 45.9$
6. $-25n + 3.8 + 2n = 329$
7. $2a + 1 = a + 7$
8. $3(5m - 1) = 7$
9. $2(x + 1) = 5(x - 5)$

ACTIVITY: Models of many forms

Eight accountants are hired to examine a company's costs and revenues. They each come up with a different equation to determine the relationship between cost and revenue. Most of the equations are correct, but not all. Determine which equations are correct models, and which are incorrect. Include all your reasoning in a well-organized manner.

1. $C(R + 5) + 9R(R - 1) = 160 + C^2 + 9R^2$

2. $C = 1.8R + 32$

3. $5C - 9R - 160 = 0$

4. $\frac{C}{9} - \frac{R}{5} + \frac{32}{9} = 0$

5. $5(C + 4) = 9(20 + R)$

6. $\frac{5}{C} - 212 = 9(R - 100)$

7. $\frac{C}{5} - \frac{R}{9} = 32$

8. $R = \frac{5}{9}C - \frac{160}{9}$

Can all problems be represented by graphs?

Each of the bigger scatterplots below has a recognizable pattern. Imagine a line of best fit for each. Why is this difficult in graphs d and e? Do you see a pattern in these graphs?

As we can see in the smaller graphs, graphs a to c have **linear** patterns. Graphs d and e have other predictable patterns, and we may draw **non-linear curves** to represent these.

Each of these curves and lines helps us to model a trend that can aid our decision-making.

ACTIVITY: How do I know it's linear?

■ ATL

■ Creative-thinking skills: Make guesses, ask 'what if' questions and generate testable hypotheses

We have seen models of patterns in the forms of graphs **or** equations. Now we are going to look at the equations for each of the graphs in the examples a to e. If we were to make a table of values and graph these equations, they would come out looking like the line or curve of best fit in each case.

a $y = 2.50x$ c $y = -0.3x + 75$ e $y = e^{0.75x}$
b $y = 15 + 0.1x$ d $y = -x^2 + 1$

What do the linear equations have in common? Can you come up with a general rule to describe what linear equations look like algebraically? Use graphing software to verify and justify your solution. Alternatively, draw a table of values and graph them by hand.

◆ Assessment opportunities

◆ In this activity you have practised skills that can be assessed using Criterion B: Investigating patterns.

Graphs **Graphs with best-fit models**

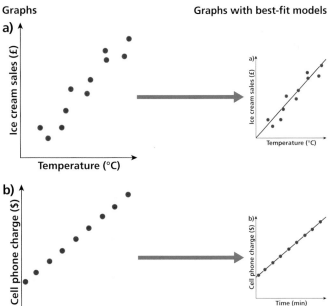

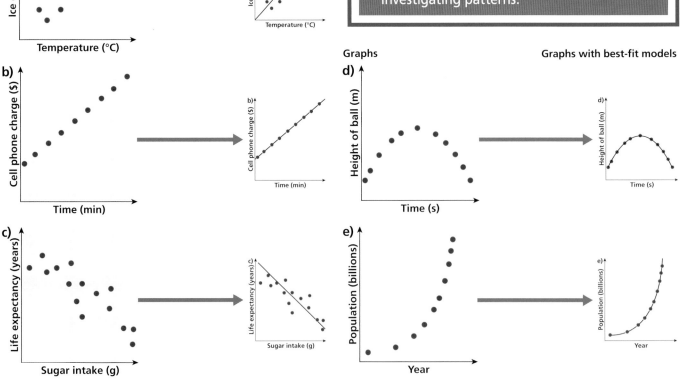

How do I determine a rate of change?

You may have noticed from the previous activity that linear equations always have just one variable whose highest power is 1, and always have a coefficient (although sometimes this is just 1). Before we look closer at these algebraic patterns in **linear equations**, let's examine their associated **graphs** in more detail, to look at where any patterns are coming from and to think about whether they accurately represent real life. You may remember this from Chapter 1 of *Sciences for the IB MYP 3: by Concept*. To begin, let's use a simple example equation and make whatever observations we can about the graphical representation.

Someone is driving at a constant speed of 60 km/h along Yongue Street – let's graph it: distance versus time.

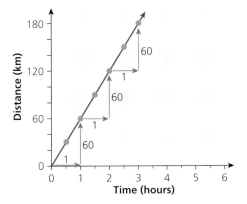

Notice that on the graph, for every one hour driven, the increase in distance is always 60 km. This means the rate of change is **constant** – implying the driver never slows down or speeds up in that time. In reality, this is probably not possible, due to outside variables such as traffic flow, traffic signals and changes in speed limits. Even if the graph looked more like the one shown top right (scattered with line of best fit) the general trend is still consistent.

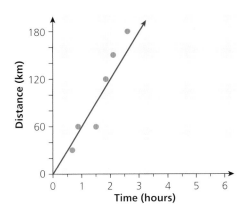

Pick any two points on the line and draw a triangle. Label the 'length' of the base and height in units. You can determine these lengths in one of two ways: skip count or calculation.

Skip count: The length of the height (vertical side), known as the **rise**, can be determined by counting from the bottom of the triangle to the top. Pay close attention to the scale when you do this – this scale increases in jumps of 20 s, so we need to count in increments of 20 s.

The length of the base (horizontal side), known as the **run**, can be determined by counting from the left-most point to the right.

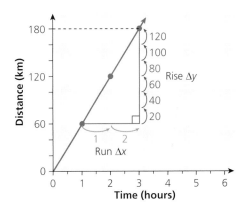

Calculation: Alternatively, we can simply read the values off the graph and calculate the differences in the top and bottom values, and the differences in the right and left values as shown on page 38.

DISCUSS

The reason the letter m was chosen is unknown – can you think of any possible connections between the letter m and the slope? What are some theories on the internet?

EQUATION

Can you think of an equation that would relate the x-values with the y-values in the table of values at bottom left? For every increase in x, y increases 60 times more. So if you multiply x by 60, you'll get y. Thus: $y = 60x$.

Notice the placement of the rate of change in that equation? Let's try a few other examples. If the car were driving at 40 km/h or 100 km/h, the tables of values would look like these:

x Time (h)	y Distance (km)
0	0
1	40
2	80
3	120
4	160

x Time (h)	y Distance (km)
0	0
1	100
2	200
3	300
4	400

What would the appropriate equations for y be in these scenarios?

$$y = 40x \qquad \text{and} \qquad y = 100x$$

Thus, the rate of change in a linear equation (where y is isolated) is the coefficient of x. We will elaborate more on the 'where y is isolated' part in upcoming sections. Mathematicians across the world use the letter m as a short form to represent the slope. Rather than constantly referring to it as 'the coefficient of x in a linear equation where y is isolated', they simply refer to it as 'm'.

We have just uncovered four different ways to determine the rate of change when it is constant. Ensure you understand the summary before moving on to the practice exercise.

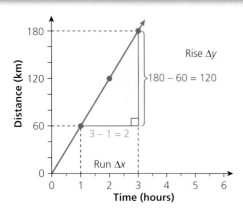

Either way, we find:
- rise = 120
- run = 2

We get this result every time, no matter which two points we select on the line to construct the triangle. So, how do these numbers (120 and 2) relate to the speed of the car? Can we use an operation between them to come up with the number 60? Yes, $120 \div 2 = 60$! This is a phenomenon unique to linear graphs – it will never be this consistent in any other graph type!

To determine the rate of change from a graph, draw a rate triangle and calculate the rise divided by the run.

TABLE OF VALUES

Let's look at a table of values for the same graph.

x Time (h)	y Distance (km)	First differences
0	0	
1	60	$60 - 0 = 60$
2	120	$120 - 60 = 60$
3	180	
4	240	

What do you notice about the y-values in this table? How does this relate to the graph? To the calculated rate of change?

ACTIVITY: Let's hit the slopes!

■ ATL

- Critical thinking skills: Draw reasonable conclusions and generalizations; Test generalizations and conclusions; Use models and simulations to explore complex systems and issues

We are going to investigate what happens to the line as the slope changes.

The general equation for calculating cost is price times quantity bought. Suppose you want to buy a few pieces of a 5 euro item. If x represents the quantity you buy, then your cost will be $5x$.

1 Draw the linear equation $y = 5x$ on your preferred graphing software (for example Desmos, Geogebra or a graphing calculator).

2 Investigate what happens to this line as you change the price in the equation. Try using big and small numbers, fractions and negative numbers. Employ as much variety as you can.

3 Clearly state any patterns you notice.

4 Verify and justify your conclusions using additional graphs.

EXTENSION

Can you make your line perfectly horizontal? Or vertical?

◆ Assessment opportunities

◆ In this activity you have practised skills that can be assessed using Criterion B: Investigating patterns.

How to find the rate of change

Slope summary

Given	How to find rate	Example
Word problem	Look for 'per' in the units, or 'for every increase in (a quantity) there is an increase/decrease in (a related quantity)'	km per h (km/h)
Graph or any two points	Use a rate triangle and calculate the rise divided by the run	Rise Δy Run Δx
Table of values	Calculate first differences	$\begin{array}{c|c} x & y \\ \hline 0 & 10 \\ 2 & 18 \\ 4 & 26 \\ 6 & 34 \\ 8 & 42 \end{array}$ $+2$ each, $+8$ each; $\Delta x = 2$, $\Delta y = 8$; slope $= \frac{\Delta y}{\Delta x} = \frac{8}{2} = 4$
Equation	Coefficient of x	$y = 3x$

THINK-PAIR-SHARE

How can you determine the slope (or gradient) of a line if given only two points on the line? Share your thoughts with a partner and then discuss as a class.

What is a fixed cost?

Ezekiel plans to make extra money mowing lawns in the neighbourhood throughout the summer. He buys a lawnmower for $200 and charges $5/hr. If we let h represent the number of hours Ezekiel works, his earnings would be $5h$ (5 dollars multiplied by the number of hours he works). His **profit** is the amount of money he earns after covering the cost of the lawnmower. This can be modelled by the equation $p = 5h - 200$, where p represents the amount of profit. Let's graph this:

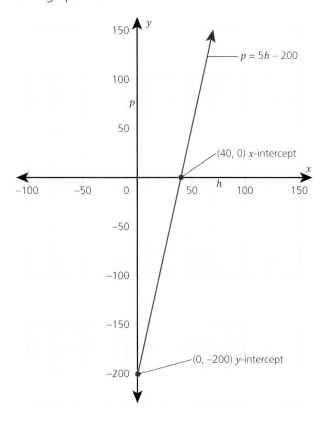

Let's consider point $(0, -200)$. Conceptually, this means $h = 0$, so no hours worked yet, and $p = -200$, meaning Ezekiel is $200 down after buying the mower. Graphically, this is the **y-intercept** – the point where his profit line meets or crosses the y-axis. In a linear graph, this only ever happens once. It is an important point as it is often thought of as the 'starting point', the point when the independent variable is zero.

What happens at the point $(40, 0)$? Here, Ezekiel will have worked 40 hours and made no profit! How is that possible? Ezekiel has covered the cost of the mower and has reached his break-even point – the point at which his costs and his earnings are equal. No profit, no loss. Notice this point crosses the x-axis. The point at which a line crosses the x-axis is referred to as the **x-intercept**.

In general, an intercept is the point where a line (or curve) crosses an axis, and at least one of the coordinates is zero. We will focus on the y-intercept at this time, and revisit the x-intercept later in this chapter.

Notice that in Ezekiel's example, the y-intercept represented a fixed cost – a cost that is a constant number, unaffected by sales or performance. In other examples, it may represent an initial temperature, initial distance from home, or other starting point of some kind. When there is no fixed cost or initial value, the two variables are directly related – they have a **direct variation**. If, however, the y-intercept is found anywhere but the origin, the variables have an **indirect variation**. We are now able to locate this initial value on a graph and to identify it in the description of a problem, but can we determine it by simply looking at an equation?

ACTIVITY: Pizza graph

■ ATL

■ Critical-thinking skills: Draw reasonable conclusions and generalizations; Test generalizations and conclusions; Analyse complex concepts and projects into their constituent parts and synthesize them to create new understanding

Since we know that the initial value gives us the starting point or y-intercept, let's use graphs to help us find the y-intercept in an equation. If a plain pizza costs C\$15 plus C\$1.50 per topping, and x represents the number of toppings, then the equation for the total cost, y, is:

$y = 15 + 1.5x$ or $y = 1.5x + 15$

Graph this using a technology of your choice, and then play with the initial cost. What if a plain pizza were C\$20 or C\$10, and so on? Continue until you are able to identify the y-intercept from an equation without a graph. Remember to verify and justify your findings.

◆ Assessment opportunities

◆ In this activity you have practised skills that can be assessed using Criterion B: Investigating patterns.

How to find the initial value

y-intercept summary

Given	How to find rate	Example
Word problem	Look for a starting point or fixed cost	In a single hour, Nick currently earns a base wage of 10 euro per hour plus 10% of each sale he makes
Graph or any two points	Where the line crosses the y-axis	
Table of values	Where $x = 0$	<table below>
Equation	Constant	$y = 2x - 13$

x	y
−2	−30
−1	−20
0	−10
1	0
2	10

PRACTICE EXERCISE

Locate the y-intercept in each case.

1

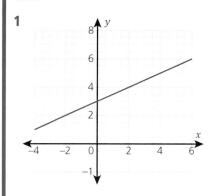

2

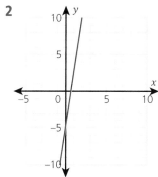

3

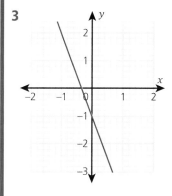

4

x	y
−5	0
0	2
5	4
10	6

Hint

Although in question 4 there is no graph from which to locate a crossing point, recall that to find the y-intercept, x must be zero.

Why do we use slope-intercept form?

$$y = mx + c$$

where m = slope and c = y-intercept.

We have been modelling problems using graphs based on tables of values. In order to fully understand what you are looking at, it is really helpful to know how to draw a graph manually. However, there are much quicker ways to draw graphs. If you know how to graph with technology, you can spend more time on the analysis.

SHORTEST DISTANCE

The shortest distance between two points is always a straight line! Thus, given the coordinates of any two points we can draw a straight line and determine its equation.

Example

Graph the line that passes through the points (–1, 1) and (3, 2).

Solution

We find the points (–1, 1) and (3, 2), plot them, then draw a line that goes through both!

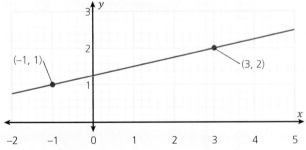

From this, if we wanted to, we could draw a rate triangle to determine Δx and Δy and then the slope. We could also see that the y-intercept is about 1.25.

Example

Find the equation of the line that passes through the points (3, 4) and (2, 7) **without graphing**.

Solution

Just like when we are graphing, to determine the slope, we need to find Δy (the change in y-values) and Δx (the change in x-values). Let's look at those points:

(3, **4**) and (2, **7**)

$$\Delta y = 7 - 4 = 3$$

(This is equivalent to finding the vertical side of the rate triangle.)

(**3**, 4) and (**2**, 7)

$$\Delta x = 2 - 3 = -1$$

(This is equivalent to finding the horizontal side of the rate triangle.)

So, $m = \frac{\Delta y}{\Delta x} = \frac{3}{-1} = -3$

Let's fill in the part of the equation we now know: $y = -3x + c$

To solve for c, we need c to be the only unknown in the equation – what can we use for x and y? Remember that x and y represent the coordinates of any point on the line. We happen to have been given **two** points on the line! We may choose either one! Let's use (3, 4) as these are easy numbers to work with. Then:

$$y = -3x + c$$

$$4 = -3(3) + c \quad \text{(Since the point (3, 4) gives us } x = 3 \text{ and } y = 4)$$

$$4 + 9 = c$$

$$c = 13$$

Thus, $y = -3x + 13$

Let's make sure this works. Let's substitute the x-coordinate from the other given point and see if we get the matching y-coordinate:

$$y = -3x + 13$$

$$y = -3(2) + 13 \quad \text{(substituting } x = 2 \text{ from the point (2, 7), hoping we will end up with } y = 7; \text{ and substituting c} = 13, \text{ which we just calculated)}$$

$$y = -6 + 13$$

$$y = 7 \quad \text{Success!}$$

SLOPE-INTERCEPT FORM

If given an equation, as opposed to a pair of points, we are essentially being given a point and directions to the next point.

directions to second point

$$y = \frac{1}{2}x + 3 \longleftarrow \text{starting point}$$

We can draw the graph, labelling the starting point and steps to the next point.

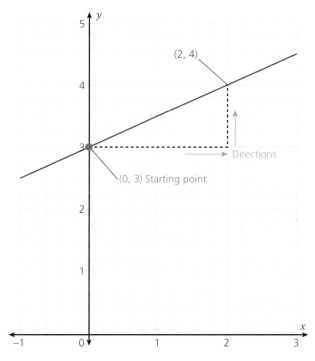

A car rental agency charges a flat rate of £24.99 to rent a standard-size car, plus £0.25 per kilometre. To help customers understand the charges, the owner, Dara, graphs what possible charges can look like. She has difficulty coming up with an equation, so begins with a table of values.

x Distance driven (km)	y Rental cost (£)
0	0.25(0) + 24.99 = 24.99
10	0.25(10) + 24.99 =
20	0.25(20) + 24.99 =
x	y = 0.25x + 24.99

Dara realizes she could have identified £24.99 as the **initial cost** (the y-intercept) and 0.25 as the **rate of change** (slope), coming up directly with $y = 0.25x + 24.99$.

Next she uses Desmos to graph her equation (page 44).

ACTIVITY: Hands-on linear equations

■ ATL

- Transfer skills: Apply skills and knowledge in unfamiliar situations
- Critical-thinking skills: Use models and simulations to explore complex systems and issues

Find a set of identical, stackable pieces, for example building blocks or paper cups.

1 **Draw and label a Cartesian plane with number of units (cups or blocks) along the x-axis, and height along the y-axis.**
2 **Measure the height of a single unit. Plot your findings on your graph.**
3 **Repeat with two units. Then with more until you have used at least five units.**
4 **Describe any patterns you see.**
5 **How does the y-intercept relate to your study? How does the slope?**
6 **Write down an equation of the line that represents the height of x units.**
7 **Verify and justify this equation.**
8 **Predict the height if you had 30 units (cups or blocks).**

◆ Assessment opportunities

- ◆ In this activity you have practised skills that can be assessed using Criterion B: Investigating patterns.

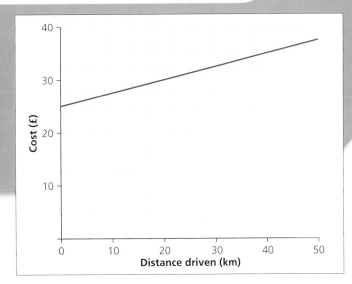

A customer posts an online review claiming that they were overcharged, having driven 30 km and paying £30. Are they justified in this claim? We can look into this one of two ways: graphically or algebraically. Let's do both!

Graphical solution

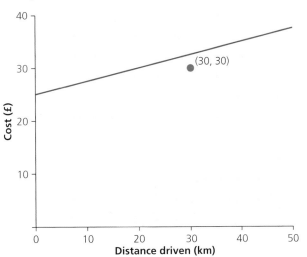

The point (30, 30) is not on the line. Looking at the graph the customer actually paid less than she should have!

Algebraic solution

If the customer drove 30 km and we substitute this number into our equation, we should get $y = 30$ (a rental cost of £30) to justify their claim.

$y = 0.25x + 24.99$

$= 0.25(30) + 24.99$

$= 7.50 + 24.99$

$= 32.49$

The cost of a rental including 30 km distance is £32.49.

Summary

To determine whether a point is on a line (whether the suggested relationship between x and y exists), we can either graph the line and see if the point lies on it, or substitute one coordinate into the equation to see if the other comes out as expected.

ACTIVITY: An entire career in one equation

■ ATL

- Information literacy skills: Present information in a variety of platforms and formats
- Transfer skills: Apply skills and knowledge in unfamiliar situations; Inquire in different contexts to gain a different perspective

A professional footballer plays for a major league team for $2 million per year and gets a one-time signing bonus of $3 million.

1 Represent the possible time and earnings combinations in:
   a a table of values   c graphic form.
   b algebraic form
2 Identify the slope and y-intercept in each case, describing how you found or calculated them.
3 a What would the athlete's total earnings be if they play for this team for eight years? Which method (algebraic or graphic) is the most helpful to you in answering this question?
   b Is the point (12, 27) on your line? (Use your equation.)

◆ Assessment opportunities

- In this activity you have practised skills that are assessed using Criterion A: Knowing and understanding.

Again, we reach a different, higher number! So the customer was right to say they had paid the wrong price – but perhaps they shouldn't have been so angry about it! Dara was able to post her findings in response to the online review.

Can a constant be graphed?

We have been examining equations for which there are always two variables. What if we eliminate a variable? What if we get rid of mx from the equation for the line and just write $y = c$?

Example

Graph $y = 5$.

Solution

We know how to find a starting point and use the slope to find a second point.

Starting point (y-intercept) = 5

But what is the value of the slope?

Let's try a table of values.

x	$y = 5$
−2	5
−1	5
0	5
1	5
2	5
3 000 000	5

It appears that no matter what the x-value is, y will always be 5. So, x can be any number from −∞ to ∞ but y will remain at 5.

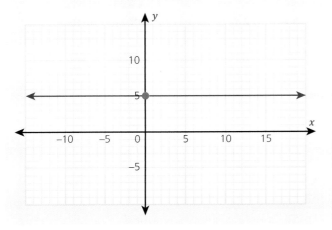

Now let's consider that slope again. We see that the rise is zero and the run is any number. Since the rise

– the numerator in the slope – is zero, the slope itself must be zero, as zero divided by anything is zero.

This explains what happened to the 'mx' part of $y = mx + c$:

$$y = 0x + c$$

$$y = 0 + c \qquad \text{(0 times anything is still 0)}$$

$$y = c$$

What if, instead of y being a constant, we have $x = c$?

Example

Graph $x = −2$.

Solution

It's difficult to find a slope or y-intercept, so let's go back to a table of values again.

$x = −2$	y
−2	−1
−2	0
−2	1
−2	2.48737821

Once again, we have a situation whereby one variable is the same and the other can be any number from −∞ to ∞. This time, it is x that is constant. Graphing these points we get:

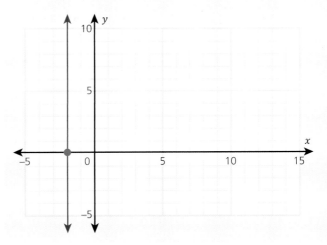

The line in the graph is a vertical line – how can we describe the slope?

rise = ∞

run = 0

slope = $\dfrac{\text{rise}}{\text{run}} = \dfrac{\infty}{0}$

But we can't divide anything by 0, so we say the slope is **undefined**.

In conclusion, when we graph constants, the resulting lines are counterintuitive!

- $y = c$ which is a horizontal line and $m = 0$
- $x = c$ which is a vertical line and m is undefined

Hint

It is a very common misconception that $y = c$ yields a vertical line, because y is the vertical axis. When you find yourself graphing a constant, take your time and, without necessarily writing it out, think through a possible table of values. The equation $y = c$ means that 'no matter what x is, y is always the same', meaning x goes from $-\infty$ to ∞ so the line stretches across horizontally.

ACTIVITY: Playing with linear graphs

■ ATL

- Creative-thinking skills: Make unexpected or unusual connections between objects and/or ideas
- Transfer skills: Apply skills and knowledge in unfamiliar situations
- Organization skills: Select and use technology effectively and productively

Part A

1 Write four equations of graphs that have the same slope (m) but different y-intercepts.
2 Graph them all on the same Cartesian plane. Use graphing software if possible.
3 Describe any patterns you notice. Can you make a general statement about equations with the same slope?
4 Verify and justify your statement.

Part B

Investigate positive and negative slopes. Describe, verify and justify any patterns or relationships you notice.

Part C

1 Using graphing software, graph a line with a positive slope.
2 Experiment with a second equation, trying to graph a line that is perpendicular to the first. (Think of the direction this perpendicular line would need to go in, and what you need to do to the slope to ensure it will go in this direction.)
3 Describe any relationship you notice about slopes of lines that are perpendicular to one another, and verify and justify your statement.

◆ Assessment opportunities

◆ In this activity you have practised skills that can be assessed using Criterion B: Investigating patterns.

Are there limits to what we can do?

CONSTRAINTS

What is wrong with this graph?

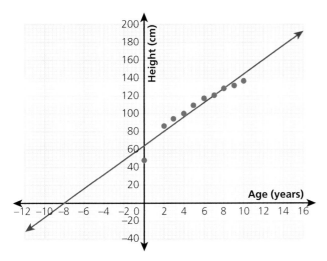

■ Average female growth trend by age

Source: The World Health Organization (WHO), Gerontology Research Center (National Institutes of Health, USA), and the U.S. Centers for Disease Control and Prevention (CDC)

Can a person have a negative height? Or be less than zero years old? Do we grow endlessly, forever?

PRACTICE EXERCISE

Suggest constraints on the following relationships:
- **temperature and electricity costs in Siberia**
- **cost of milk over a period of years**
- **a country's debt over a period of years**
- **sales and number of cold calls (unrequested marketing phone calls)**
- **number of songs downloaded and amount left on music gift card**
- **cost of a magician at a children's party and the number of children present**
- **taxi fare and journey distance.**

Make sure you don't graph your equation with its arrows, suggesting it goes on forever, when the context of the problem suggests otherwise. A more appropriate graph would look like this.

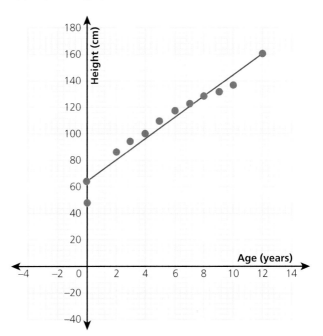

This graph reflects the fact that ages and heights are always positive numbers. What would this look like algebraically?

$$y = mx + c \qquad\qquad y > 0 \qquad\qquad 0 < x < 12$$

The expression $y > 0$ is called a **constraint**. It limits the values y can take on. In this case, it limits them to values greater than zero (that is, positive numbers).

There is also a constraint on x, which is between 0 and 12. This tells us that at age 12, female growth stops or slows down enough to change the shape of the graph.

! Alone or in groups of up to three, select an existing charity or community project you would like to support. Decide on a specific goal for this charity – for example, purchasing 200 turkeys for a soup kitchen or 10 new beds for a shelter – and find out how much money you will need to raise in order to meet this goal.

1 Come up with three different fundraising activities (for example, bake sale or walkathon).

2 For each fundraising activity, provide a detailed table of the costs of all your materials and any other start-up expenses, and develop an equation to determine the amount that can be raised after expenses.

3 Graph your three equations – which has the steepest slope? What does that mean?

4 Carry out the activity you found to be the most effective for raising funds, and inform your donors of your results!

5 Reflect on how you could develop this activity on a larger scale to reach more people and climb higher on your graph.

MEET A MATHEMATICIAN: STEPHEN HAWKING (1942–2018)

Learner Profile: Knowledgeable, Reflective

Many of the world's greatest minds have experienced harsh constraints and still excelled beyond what most of us could hope to achieve. Stephen Hawking was diagnosed with amyotrophic lateral sclerosis (ALS) at the age of 21. Hawking viewed his physical 'constraint' as something that drove him to deeper reflection and understanding:

'Before my condition was diagnosed, I had been very bored with life,' he said. 'There had not seemed to be anything worth doing.' Indeed he averaged one hour per day on his studies. Doctors predicted that he would not live long enough to complete his PhD. This prompted a change in attitude: 'I therefore started working for the first time in my life. To my surprise I found I liked it.'

Hawking went on to change the way people think about black holes. Black holes are sites of gravity so powerful that it was believed nothing could escape from them – but, using mathematics, Hawking was able to justify that black holes actually emit a form of radiation. Based on this, he suggested the possibility that an object entering a black hole could be released into another universe, perhaps in another form. The young student who had been given two years to live was eventually awarded one of the most prestigious academic posts in the world – Lucasian Professor of Applied Mathematics at Cambridge University – a post he held for 30 years.

Hawking's greatest achievements relate to his contributions to Cosmology and Relativity, but his unique presence in popular culture came from his ability to describe complex theories to ordinary people. In his book, *A Brief History of Time*, he put a case for time travel and suggested the possibility for humans to colonize other planets. He authored 14 other books, and appeared in cartoons, comedies and sci-fi programmes. His early life is documented in the critically acclaimed film, *The Theory of Everything*.

Hawking also focused on global fairness and development. He was given a $12 million grant to study the negative effects of **too much** development – of what artificial intelligence could do to compromise our security and economy. He also advocated for colonizing space … just in case. He turned down lecture invitations to countries that he felt were not acting fairly.

When is standard form a better alternative?

TIME TO FREEZE

Olivera filled an ice-cube tray with water and placed it in the freezer. Once the tray was in the freezer, she turned on her timer. At regular intervals, she recorded the temperature of the water as it turned to ice. The general trend is shown in the graph.

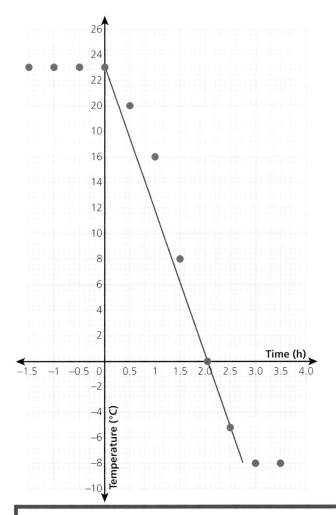

DISCUSS

What was the temperature of the water when the timer started? Why doesn't the red line start at the origin?

What might 'negative time' represent in this case?

When did the water turn to ice (that is, reach 0°C)?

When the timer started, time = 0. The y-intercept is found at $t = 0$, and represents the initial temperature of the water. If the graph had started at the origin, it would mean the water was already solid ice before Olivera began her experiment!

In this scenario, 'negative time' is the time before Olivera started her timer – notice the water didn't change temperature before the experiment started. It takes two hours for the water to become ice at a temperature of 0°C – this is the x-intercept.

TICKET COMBINATIONS

Suppose a movie theatre charges $20 per regular (child or adult) ticket and $15 per senior ticket. The manager wants to determine what possible regular–senior combinations will earn $120. If R represents every regular ticket and S represents every senior ticket, then the total earnings for the theatre would be:

$$20R + 15S = 120$$

This equation is not in a form we are used to graphing! The variables here are R and S. We are used to thinking about x and y, and graphing on the x- and y-axes. We can continue with R and S, and create identical R- and S-axes, but to keep things familiar, let's redefine the variables:

$$20R + 15S = 120 \quad \text{becomes} \quad 20x + 15y = 120$$

where x is number of regular tickets and y is number of senior tickets sold. However, this equation is still not in a form that gives us slope and intercept.

To rearrange, isolating y:

$$15y = 120 - 20x$$

$$y = \frac{120 - 20x}{15}$$

$$y = 8 - \frac{20}{15}x$$

$$y = 8 - \frac{4}{3}x$$

Drawing up a table of values is time-consuming and risks introducing errors. We could just find any two points on the line and connect them. Which are the easiest points to find? The x- and y-intercepts, because each involves substituting the easiest number to work with, zero!

Step	x-intercept	y-intercept
	(Let $y = 0$)	(Let $x = 0$)
	Circle this before moving on to next step!	Circle this before moving on to next step!
1	$20x + 15(0) = 120$	$20(0) + 15y = 120$
2	$20x = 120$	$15y = 120$
3	$x = \frac{120}{20}$	$y = \frac{120}{15}$
4	$x = ⑥$	$y = ⑧$
	(6, 0) is the x-intercept	(0, 8) is the y-intercept

Setting one coordinate to zero effectively cancels out an entire term from the equation, so the process of finding intercepts is quick and easy.

This new form of linear equation is known as **standard form**. Once the intercepts are determined, we can graph the equation.

Now we can graph, use trial and error, or a TOV.

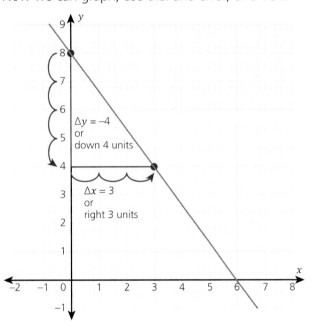

■ Every point on this line represents a possible regular–senior ticket combination that would earn $120. Some are purely theoretical – you can't buy half a ticket!

If we preferred, we could use a table of values to determine possible ticket combinations.

x	$y = 8 - \frac{4}{3}x$	
0	$y = 8 - 0 = 8$	
1	$y = 8 - \frac{4}{3}(1) = 6\frac{1}{3}$	Can't buy $\frac{1}{3}$ of a ticket!
2	$y = 8 - \frac{4}{3}(2) = 5\frac{1}{3}$	Also, not possible
3	$y = 8 - \frac{4}{3}(3) = 4$	A whole number of tickets!

Notice that only x-values which are multiples of three give integer answers. So, we can make life a bit easier:

x	$y = 8 - \frac{4}{3}x$
0	$y = 8 - 0 = 8$
3	$y = 8 - \frac{4}{3}(3) = 4$
6	$y = 8 - \frac{4}{3}(6) = 0$
9	$y = 8 - \frac{4}{3}(9) = -4$

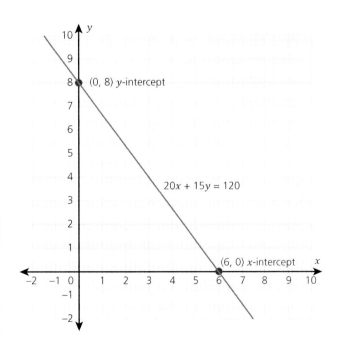

ACTIVITY: What did Marko do wrong?

Marko's teacher set this question: Find two points on the line $20x + 15y = 120$. Here is Marko's solution:

x-intercept	y-intercept
Let $y = 0$	Let $x = 0$
$20x + 15(0) = 120$	$20(0) + 15y = 120$
$20x = 120$	$15y = 120$
$x = \frac{120}{20}$	$y = \frac{120}{15}$
$x = 6$	$y = 8$

Marko says he has found point (6, 8). Has he?

It is a very common mistake to combine the x-intercept x-coordinate and the y-intercept y-coordinate into one single point. But an intercept must always have zero as one coordinate.

PRACTICE EXERCISE

1 Identify the point that is **not** on the line $15x + 60y = 180$ (or find the value that is not true for the equation).

   a (4, 2)   c (12, 1)

   b (0, 3)   d (8, 1)

2 Isolate y in each equation (write y in terms of x).

   a $6x + 2y = 24$

   b $42x - 14y = 21$

3 Determine the x- and y-intercepts for each equation in question 2.

4 A restaurant employs seven servers and three bartenders.

   a Write an equation that shows all wage combinations resulting in a $100 per hour payroll.

   b Graph your equation.

   c How much would the bartenders earn per hour if the servers earn $10? Use your graph and your equation to justify your response.

ACTIVITY: Candy combos

A bulk-food store is making a mix of cashews and chocolates. The cashews cost A$30/kg and the chocolates are A$8/kg. The store will spend A$150 on the mixture. Represent the possible cashew–chocolate combinations in:

1 a table of values
2 algebraic form
3 graphic form.

Identify the x- and y-intercept in each. What do they mean in the context of the scenario?

If the store buys 4kg of cashews, can they buy 2kg of chocolates?

How can linear equations influence our choices?

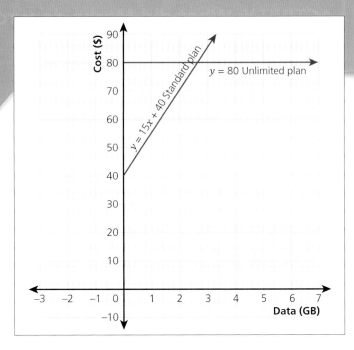

A mobile-phone company has the following plans on their website.

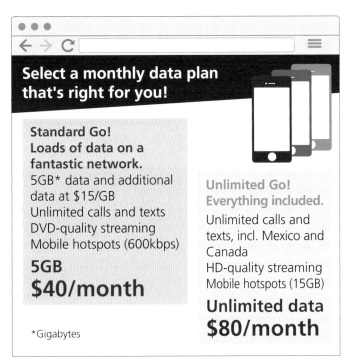

Select a monthly data plan that's right for you!

Standard Go!
Loads of data on a fantastic network.
5GB* data and additional data at $15/GB
Unlimited calls and texts
DVD-quality streaming
Mobile hotspots (600kbps)

5GB
$40/month

*Gigabytes

Unlimited Go!
Everything included.
Unlimited calls and texts, incl. Mexico and Canada
HD-quality streaming
Mobile hotspots (15GB)

Unlimited data
$80/month

Which plan should you choose? Let's graph it!

Standard plan: $40 for 5 GB (there's our starting point!), and $15/GB thereafter ('per GB' tells us this is a rate of change. Think slope!).

$$y = mx + c$$

$$y = 15x + 40$$ where y represents the total monthly cost and x represents the number of GB used **over 5 GB**.

Unlimited plan: $80. Always, with no additional charges. If y represents the total monthly cost, then $y = 80$ (because the cost never changes).

Graphing, we get the lines shown on the top right:

Examples

1   When is the unlimited line higher than the standard line? What does this mean?

2   When is the standard line higher than the unlimited line? What does this mean?

3   At what point do the two lines intersect? What is the significance of this point?

Solutions

1   The unlimited line is higher from $x = 0$ until $x = 2.67$ (roughly). This means that even if the customer uses an additional 2 GB over the 5 GB allowance, it is still cheaper than the unlimited plan.

2   The 5 GB line is higher than the unlimited line after $x = 2.67$. This means that if the customer uses more than an additional 2.67 GB, the unlimited plan is actually cheaper.

3   The two plans intersect at (2.67, 80) – this is the point at which they cost the same.

> **DISCUSS**
>
> Which plan is the better choice? What must be true of the slopes for two lines that **never** intersect?

SUMMATIVE ASSESSMENT:
Fair pay

Use these problems to apply and extend your learning in this chapter. The problems are designed so that you can evaluate your learning at different levels of achievement in Criterion D: Applying mathematics in real-life contexts.

Andrea earns $8.75/hour washing dishes at a restaurant. Brett waits tables at the same restaurant and earns $100 per week plus tips; tips average out to be $4.75/hour. Neither Andrea nor Brett work more than 30 hours in a single week.

1   Write an equation for Andrea's earnings based on the number of hours she works. Be sure to define your variables.
2   Now repeat Q1 for Brett's earnings.
3   Copy and complete the table of values for Andrea, using the third column to calculate first differences. What conclusions can you draw about the graph relationship?

Number of hours worked	Andrea's earnings	First differences
0		
10		
20		
30		

4   Construct a table of values for Brett.
5   Graph and identify Andrea's and Brett's equations on a suitable grid. Include all necessary labels.
6   State the y-intercept for Brett's model. What does this mean in the context of the problem? Would Brett's boss be happy with this arrangement? Why or why not?
7   State the value of the slope in Andrea's model and explain its meaning in the context of the problem.
8   State the approximate coordinates of the point of intersection of the two lines. Describe the meaning of this point with respect to the hours worked **and** the money earned.
9   If Andrea and Brett each work a 15-hour week, who will make more money? By how much?
10 Use equations to predict Andrea's and Brett's earnings if they were each to work a 40-hour week. Show all calculations.

Reflection

Questions we asked	Answers we found	Any further questions now?
Factual: How do I know it's linear? How do I determine a rate of change? What is a fixed cost?		
Conceptual: Can different expressions be equal? What are the different ways of solving an equation? Can a constant be graphed? How can linear equations influence our choices?		
Debatable: Can all problems be represented by graphs? When is standard form a better alternative?		

Approaches to Learning you used in this chapter:	Description – what new skills did you learn?	How well did you master the skills?			
		Novice	Learner	Practitioner	Expert
Organization skills					
Critical-thinking skills					
Transfer skills					
Creative-thinking skills					
Communication skills					
Information literacy skills					

Learner Profile attribute(s)	Reflect on the importance of being knowledgeable and reflective for your learning in this chapter.
Knowledgeable	
Reflective	

3 How does a network work?

Global networks are built on **logic** and are **changing** the way we handle data, make decisions and **design models**.

YOU ARE HERE

CONSIDER THESE QUESTIONS:

Factual: What are networks? How does logic lead to graph theory? What are decision trees and how do they work? What do we mean by 'invisible algorithms'?

Conceptual: How can games be based on logic? Can puzzles be modelled? How can algorithms change things for us?

Debatable: How old are games? How can networks connect us in a global way? Are they always positive?

Now **share and compare** your thoughts and ideas with your partner, or with the whole class.

○ IN THIS CHAPTER, WE WILL ...

- **Find out** how logic can influence or change our decision-making.
- **Explore** the connection between invisible algorithms and our everyday lives.
- **Take action** by appreciating, and improving, our library and media-centre spaces.

■ These Approaches to Learning (ATL) skills will be useful ...

- Critical-thinking skills
- Media literacy skills
- Creative-thinking skills
- Affective skills
- Transfer skills

◆ Assessment opportunities in this chapter:

- **Criterion A**: Knowing and understanding
- **Criterion B**: Investigating patterns
- **Criterion C**: Communicating
- **Criterion D**: Applying mathematics in real-life contexts

■ From this …   ■ … to this!

We will reflect on these Learner Profile attributes …

- **Principled** – We act with integrity and honesty, with a strong sense of fairness and justice, and with respect for the dignity and rights of people everywhere. We take responsibility for our actions and their consequences.

- **Balanced** – We understand the importance of balancing different aspects of our lives – intellectual, physical, emotional – to achieve well-being for ourselves and others. We recognize our interdependence with other people and with the world in which we live.

THINK-PAIR-SHARE

Look at this diagram. Which words do you recognize? Which ones are new to you? What do they make you imagine? How do they make you feel? How are they all connected?

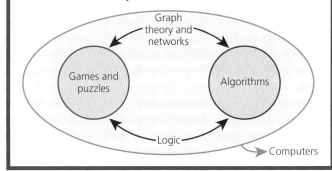

PRIOR KNOWLEDGE

Reflect on what you already know about:

- how to draw closed shapes, polygons and irregular shapes
- vertices to shapes
- how to calculate totals
- what games are and why people play them.

KEY WORDS

maze	invisible
opponent	algorithm
vertex	process
network	

How old are games?

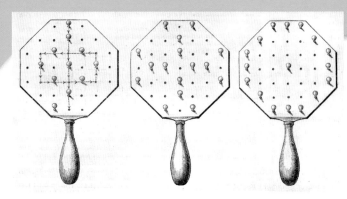

Games have probably existed for as long as humans. We have always desired to amuse ourselves and to test our own abilities, both physically and mentally. Some games, like chess and mancala, have rules and require strategy. Other games, like marbles and dice, depend on chance. Still other games require a good understanding of spatial relationships, like puzzles, or reward creative thinking, like riddles.

In this chapter, we will consider how games have given us an insight into logic, and how that logic has helped us create and understand networks. These networks impact our lives in incredible, invisible and undefinable ways.

WHAT IS PEG SOLITAIRE?

Peg solitaire, or simply solitaire, is a game for one player that may have been created by a prisoner in the Bastille in the 18th century, but there is evidence that humans have been playing similar games for far longer. The board consists of a number of bored holes, all but one filled with a marble or peg.

■ Can you make any of these final patterns, by playing the game?

ℹ️ **Did you know?**

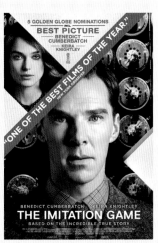

In Chapter 6 you will learn more about the mathematician pictured here, Alan Turing. He was a gifted mathematician and codebreaker whose life has inspired many books and movies, including *The Imitation Game* (2014). A letter he wrote to an eight-year-old girl (the niece of his therapist) in 1953 explained in detail how to always win at solitaire. He sent it to her as an amusement to keep her entertained while she was on a train journey to Switzerland and it contained a method, or **algorithm**, which was illustrated and coded with symbols. The letter sold for over $100 000 in 2016. View the letter online by searching for **Alan Turing solitaire letter**.

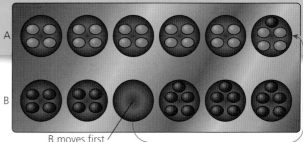

B moves first
Seeds taken from here

- Read the instructions below to find out how to play mancala.

The player moves the pegs around the board in a particular way – one peg can jump over another, if it can hop into a **vacant** (empty) space. The leap-frogged peg is removed. Jumps can only be made horizontally or vertically, never diagonally. The purpose is to remove all the pegs, landing the last one in the centre. It is not as easy as it sounds!

There are dozens of varieties of this game; some use a different starting set-up and others use variations on the rules. Why not give it a try? Remember to record how many moves it took, to see if you can beat your own record next time.

MANCALA

Mancala is an African game of strategy for two players in which each player begins with a certain number of 'seeds', represented by pebbles, shells, real seeds or beans, and a board with a series of hollowed out cups laid out in rows. Players take turns to distribute their seeds into the cups in a move called 'sowing', aiming to capture the opponent's seeds. The aim of the game is to capture as many seeds from your opponent as possible, and the player with the most seeds at the end of the game wins.

Most mancala games begin like this.

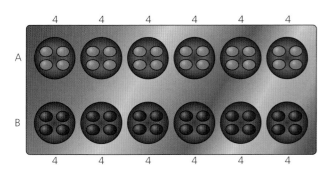

The players take turns. Alternating, they lift their seeds from one of the holes on their side of the board and sow them, one by one, into sequential holes in an anticlockwise direction.

If the last seed drops into an opponent's hole to make a total of two or three, then those seeds are captured. Rules vary, but sometimes the seeds on either side of that hole are also taken, and sometimes only seeds from **contiguous** (touching) holes with two or three pebbles are taken.

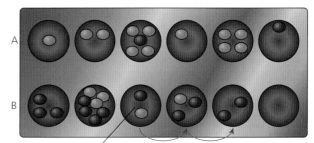

If A lands like this ... then the player captures 2 + 3 + 2 seeds

The game carries on in this way, with players capturing and recapturing the seeds. There are various rules and a variety of versions of this game, which have been played in many cultures over hundreds of years.

- This image shows a batik mancala from Java, Indonesia. The game is also known as dakon or congklak in South East Asia. Why not try to create your own mancala board and learn to play it?

MAZES

The labyrinth (life-size maze) has been a powerful and enduring form of puzzle for thousands of years. Intricate paths and winding routes have been created out of hedges, within maize fields, on chalk hillsides, cut into lonely and remote turf bogs, and traced into sands around the globe.

A labyrinth was integral to the famous Greek legend of Theseus and the Minotaur. The Celts of Northern Europe were especially fond of swirling mystical mazes, consisting of long, winding paths and dead ends. Modern mazes are often presented as amusements, places to wander in or to marvel at. They inspired the science fiction novel and movie trilogy, *The Maze Runner*, in which young people have to survive in a deadly maze that keeps changing and moving.

ACTIVITY: A-maze-ing

■ ATL

- Creative-thinking skills: Make guesses, ask 'what if' questions and generate testable hypotheses

This image shows a circular maze with only one exit.

Can you find a way from the centre to the exit on the outer edge?

Did you use the left-hand traversal rule? Why or why not?

Could the same method work using the right hand?

Consider the following scenarios. Would the left-hand transversal rule be an appropriate method to use in each situation? Explain your decision with a reason.

PRACTICE EXERCISE

Can you find the way out of this maze?

Is there a way to find your way out of the maze, a way that never fails? One method is to try to **traverse** the maze using your left hand. Traverse means to cross – in this case to navigate the maze and always find your way out. The **left-hand traversal rule** means that if you put your left hand on the wall and walk, never letting go of the wall, you will traverse the maze.

- Someone has lost their bag inside the maze and needs to find it.
- A small child is lost and alone, somewhere in the maze.
- The gardeners need to prune the maze hedge and want to start from the middle.
- Smoke is coming from somewhere in the maze.
- It is starting to drizzle and you want to get out before it rains heavily.
- The emergency services have received a call and need to get to a person who has been taken ill in the maze.

If a cave diver was lost in a cave system, would you recommend this method of solving the maze in the cave system?

■ Would you recommend using the left-hand transversal rule in this maze?

PRACTICE EXERCISE

This is a map of the personal projects on display in an exhibition hall, viewed from above.

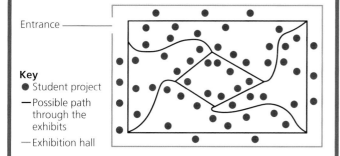

Entrance

Key
● Student project
— Possible path through the exhibits
— Exhibition hall

1 How many projects are in the hall?

2 Find one path through the maze. How many projects did you see?

3 If you used the left-hand transversal rule here, how many projects would you see and how many would you miss?

4 How many different pathways are there through this exhibition? How many decision points are there?

Other ancient games

There are many other ancient and fun games that are based on reason and logic. Pick some of the games listed below and research them. Where do they come from? Do we know? What versions are there in different places? Could you make up a game?

- Ko-No
- Fox and geese
- Tic-tac-toe
- Mu-torere
- Nine men's morris
- Whai
- Cat's cradle

How can games be based on logic?

▼ Links to: Individuals and societies

As we know from Chapter 1, the Ancient Greeks were fascinated by mathematics. They sought to use it as a tool to uncover 'truth'. They believed logic, both in a philosophical sense and a mathematical sense, was the basis and foundation of all truth. In that chapter we discussed the discoveries of the Ancient Greeks, but now it is time for us to consider how we know this information about them, passed down through the millennia.

One of the reasons we know so much about the work and teachings of the Ancient Greeks is thanks to the Muslim scholars, who preserved Greek texts from destruction by religious extremists and translated them for their own understanding. Without this dedication, we would have only a poor (and possibly no) understanding of what the Ancient Greeks knew, when and why. The destruction of wars and natural disasters claim

Now let's try some puzzles which test your number sense, logic or critical thinking.

PUZZLE A

Round 1: The goal is to place the numbers 1, 2, 3, 4, 5 in such a way that each circle makes a sum of 7. You may use each number only once.

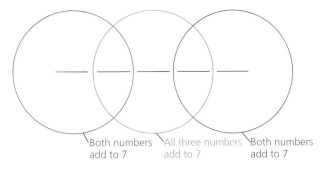

Both numbers add to 7   All three numbers add to 7   Both numbers add to 7

Round 2: Repeat the puzzle, but this time can you make the numbers in each circle add to 6?

Round 3: Can you make this puzzle work with the same numbers to give another possible total? Explain your answer with an example.

PUZZLE B

Round 1: Can you place the numbers 1 to 6 in a connected diagram such that no number is connected to a **consecutive term** (that is, to a number one larger or one smaller)? This means that 2 cannot be directly connected to 1 or 3, for example.

Where will you place the numbers here? You can use each number only once.

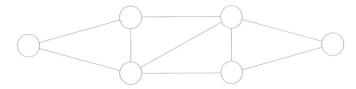

Hint

Having trouble getting started? Think strategically about which number has to go on each end.

Round 2: Repeat but for a new connected diagram and using the numbers 1 to 8, inclusive.

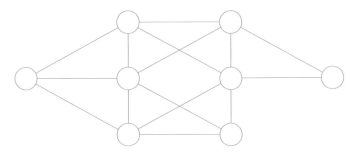

Round 3: Can you place the same numbers in this different connected (network) diagram?

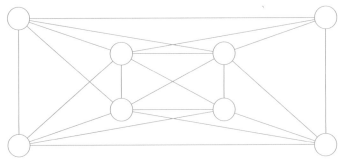

records even in modern times. Imagine what we have lost during the time from then until now!

Arguably, the most famous library of all time was the ancient Library of Alexandria in Egypt. This library was the one of the largest collections in the ancient world and also an important centre of scholarship. Its destruction is well known, a symbol of loss of knowledge and culture, because it housed hundreds of thousands of scrolls – all of which were destroyed by fire.

Diophantus was a mathematician and curator of knowledge from Alexandria. We know about Diophantus' works from the writings of Hypatia, the first known female mathematician, and the work of later Arabic mathematicians. All we really know about him personally is from his tombstone, on which the following algebra riddle was written:

'Diophantus passed $\frac{1}{6}$ of his life in childhood, $\frac{1}{12}$ in youth and $\frac{1}{7}$ more as a bachelor. Five years after his marriage was born a son who died four years before his father at half the age at which his father died.'

How old was Diophantus when he died?

Take action

The power of libraries

! Alexandria made libraries synonymous with centres of learning and academic excellence. This tradition of the library as a place of learning continues today, even in the age of easily available information. Libraries are places for calm reflection, selected and organized information, and sources of inspiration. Librarians are the curators of knowledge, lovingly keeping hold of articles until we know that we need them.

! We can become lazy in our quest for knowledge, letting search engines and their algorithms make recommendations. Try some of the following suggestions and take action with respect to your local library.

- Find out when 'Librarian's Day' is. Thank your librarian on that day!

- Check out a book. If you don't see anything that interests you, ask your librarian for a suggestion.

- Check whether the library has any of the following titles:

  - *The Imitation Game*
  - *Hidden Figures*
  - *The Man who Knew Infinity*
  - *A Mathematician's Apology*

What about other types of riddles?

PRACTICE EXERCISE

1 Change SHIP to CHIN.

2 Change ZENO to HARD.

3 Change SAND to MIND in three moves but in two different ways.

ACTIVITY: Easy does it

■ ATL

■ Creative-thinking skills: Make unexpected or unusual connections between objects and/or ideas

Lewis Carroll, author and creator of the beloved *Alice in Wonderland* books, was a mathematics professor by day. He loved riddles and logic, as we can see from his stories. He created, or popularized, logic puzzles known as 'Carroll's doublets'.

In these games, players start with one word and have to turn it into another word by changing one letter at a time. Each change must result in a new real word.

For example, we can change DOG to CAT like this: DOG → DOT → COT → CAT.

Can you create similar challenges? Make sure they are possible before you ask others to complete them.

◆ Assessment opportunities

◆ In this activity you have practised skills that are assessed using Criterion C: Communicating.

Try the *Mathematics for the IB MYP 3: by Concept* Teaching and Learning Resources for lots more of these puzzles and some fun examples of **cryptarithms** like this one.

solve my maths @solvemymaths · 14m
One of the most perfect cryptarithms ever.
Each letter is a number between 0–9.
Can you solve it?

```
  O N E
  N I N E +
T W E N T Y
  F I F T Y
E I G H T Y
```

There is a different version of Lewis Carroll's game that people play. Instead of using words, the aim of the game is to link from one celebrity to another through a series of relationships between them or movies they have starred in together. To see how this works, search for **six degrees of Kevin Bacon** or try **http://demonstrations.wolfram.com/ GraphForSixDegreesOfKevinBacon/**

Now try for yourself! Can you connect the actor Jennifer Lawrence to Liam Hemsworth through movies or TV shows they have starred in with other actors?

FROM CHANGING TO REMOVING

We are going to call on our spatial reasoning for the next riddle. How can we turn this image of four boxes into two boxes, by removing two line segments only?

MEET A MATHEMATICIAN: HANNAH FRY, BORN 21 FEBRUARY 1984

Learner Profile: Principled, Balanced

Dr Hannah Fry is a lecturer in the Mathematics of Cities at the Centre for Advanced Spatial Analysis at University College London. She works alongside a unique mix of physicists, mathematicians, computer scientists, architects and geographers to study the patterns in human behaviour, particularly in urban settings. Her research applies to a wide range of social problems and questions, from shopping and transport to urban crime, riots and terrorism.

Alongside her academic position, Hannah is an experienced public speaker, giving conferences and keynotes on the joy of mathematics, including her popular TED Talk, The Mathematics of Love. She has also published two popular mathematics books, has her own radio show, *The Curious Case of Rutherford and Fry*, and has been involved in the development of several BBC documentaries, including *How to Find Love Online* (BBC2) and *The Joy of Data* (BBC4).

You can find her on Twitter and Instagram @fryrsquared

To see how Hannah has applied the ideas of logic and mathematics to games, try the following links:

- www.telegraph.co.uk/tv/2016/07/20/a-spirited-and-fascinating-tour-of-the-information-age---the-joy/
- www.stylist.co.uk/life/how-to-win-the-annual-christmas-board-game-showdown-pictionary-monopoly-scrabble

or use Hannah Fry scrabble. The variety of her work shows her to be open-minded as to how other people can become interested in mathematics, and she is principled in sharing her mathematical ways to win at games!

Another fun type of puzzle you might have tried before is trying to join dots or draw a shape without lifting your pen. In this popular example you must connect all nine dots, using only four lines and without passing through a dot (or point) twice.

PRACTICE EXERCISE

Can you draw these images without lifting your pen from the page, that is in one continuous unbroken movement?

4 5 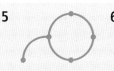 6

1 2 3 7 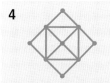 8  9

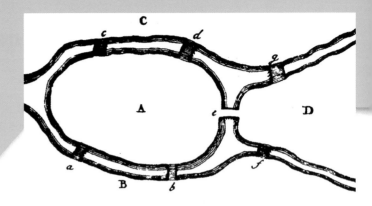

Can puzzles be modelled?

THE BRIDGES OF KÖNIGSBERG

Bridges are beautifully 'mathemagical'. Arguably, the most famous bridges in mathematics are the Bridges of Königsberg, modern day Kaliningrad. The city, on the Pregel river, consist of two islands connected to the banks, and to each other, by seven bridges. People have always enjoyed strolling across the bridges and promenading around the town. They wondered, was it possible to traverse the town crossing every bridge only once? They found it to be impossible but couldn't explain why.

Try for yourself. See if you can trace a route through the city, using every bridge only once and without repeating any stretch between them.

Later the Swiss mathematician, Leonhard Euler (whose work on face, edges and vertices we saw in *Mathematics for the IB MYP 2*, Chapter 3) became interested in this puzzle and proved mathematically why it couldn't be solved.

First, let's make the bridges, as the connections between regions, a little more obvious.

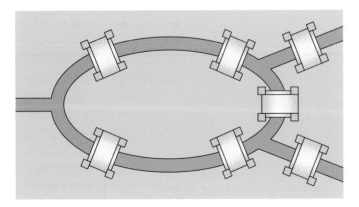

Next, as Euler did, we will imagine that each of the landmasses represents a single destination or point (a

node or a vertex) and the bridges are what connects them. We will label the landmasses A to D and the bridges 1 to 7 to avoid confusion.

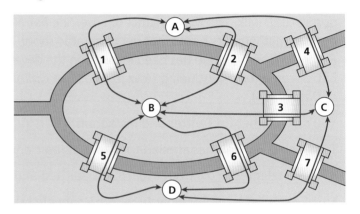

By removing the geographical information and extending the edges, we can represent the same information in a simplified way, a **network diagram**.

The power of the network diagram is its simplicity – it removes all the unwanted distraction of other visuals and detail, simplifying the map to a series of lines and dots to be counted and solved.

Euler identified whether a vertex was odd or even by counting the number of lines (edges) coming out of it. For example, D is an odd vertex because it is connected by bridges 5, 6 and 7.

We will go on to look at the significance of the numbers of odd and even vertices.

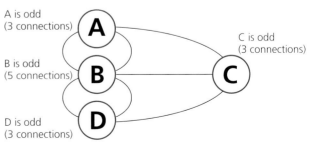

A is odd
(3 connections)

B is odd
(5 connections)

D is odd
(3 connections)

C is odd
(3 connections)

ACTIVITY: Can they be crossed or not?

ATL

- Affective skills: Mindfulness – Practise focus and concentration

Investigate these 10 networks. For each diagram, complete a new row on a copy of the table provided. Make sure you count the number of edges or lines coming from each vertex or node. Include the example of Königsberg in your table.

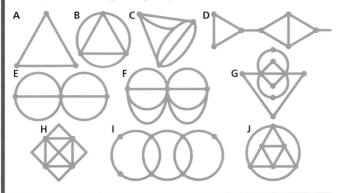

Did it make a difference if the edge was curved or straight? What patterns or general rules can you identify from the table? Prove, or verify, your rule to predict whether the following networks can be transversed.

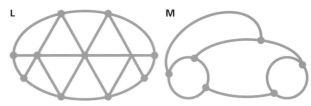

For any of the traversable or crossable networks in your table, is there a vertex that is best to start from? Identify the best starting vertex in each case, giving reasons if possible.

Can a network be traversed if it has one odd vertex only?

Can it be traversed if it has:
- **exactly two odd vertices**
- **more than two odd vertices**
- **only even edges**
- **no even vertices at all?**

Is it true that a network can be traversed if it has more odd vertices than even?

You may include drawings in your answers to test or show your reasoning.

Network	Total number of vertices	Number of even vertices	Number of odd vertices	Can it be traversed (crossed)?
A				
B				
C				
D				
E				
F				
G				
H				
I				
J				
Königsberg				

◆ Assessment opportunities

◆ In this activity you have practised skills that are assessed using Criterion B: Investigating patterns and Criterion C: Communicating.

SO WHAT DOES THAT TELL US ABOUT NETWORKS?

A network with more than two odd vertices cannot be travelled in a single trip.

There are lots of different stories about Euler's involvement in this problem, which became amazingly important in mathematics. It **could** simply have remained an interesting puzzle for Köningsberg residents but instead, through mathematical logical

thinking and abstraction, it became the fundamental basis for a whole new area of mathematics.

Watch this video to make sure you can visualize Euler's relationship and appreciate how important this puzzle is in the field of mathematics: **www.youtube.com/watch?v=nZwSo4vfw6c**

How does logic lead to graph theory?

WHAT ARE NETWORKS?

Euler's diagram is a perfect example of a simple network. Let's look at networks in more detail.

As we saw in the example of the Bridges of Königsberg, relations or connections can be modelled using a network diagram. It is useful to be able to name each part of the diagram and to understand what they show. Let's look at the vocabulary for networks, some of which we have already seen in the Königsberg investigation.

A **network** is a diagram that shows points connected to other points to indicate a relationship.

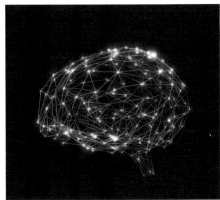

Here are the parts of a network that you will need to know:
- **Vertices or nodes** – these are points or connections on a network where something happens. In the example from Königsberg, the vertices were the landmasses. In a social network, the vertices or nodes are people or users.
- An **edge** – this is the line that connects a vertex or node to another one. It indicates a connection between them, like a road between towns or a friendship between people.
- An **arc** – this is similar to an edge but it has a **direction**, so the connection acts in one way only, like one-way streets and even some friendships!
- A **path** (or pathways) – these show how to get from one vertex to another by travelling along a series of edges or arcs.
- **Adjacent vertices** – these are vertices that are directly connected to one another.
- **Degrees of a vertex** – this tells us how many edges or arcs join at a vertex. The degrees of vertices in a network determine whether that network can be traversed (or is traversable) in one continuous line.
- **Weighting** – this describes when an edge or arc has a value associated with it, like cost, time or distance.

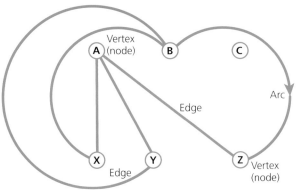

PRACTICE EXERCISE

For each of the networks shown, find the:

1 **number of vertices**
2 **number of edges**
3 **degree for each vertex (node).**

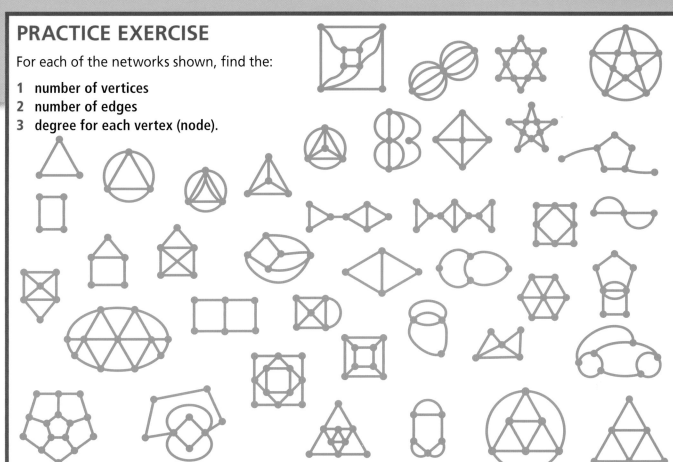

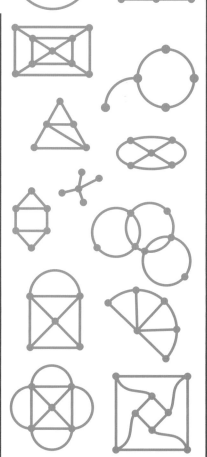

HOW DO NETWORKS USE REAL-WORLD INFORMATION?

When we are dealing with networks in real life, the vertices usually represent a noun (person, place or thing) and the edges or arcs show the relationship.

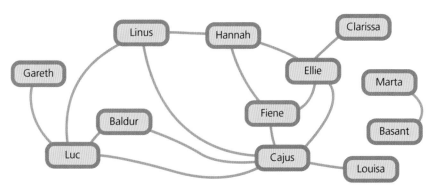

■ A friendship network in a playgroup of small children.

The purpose of this diagram is to show instantly which children play together – it is a friendship map. We can see straightaway that, as there are 12 nodes in the diagram, there are 12 kids in this group. We can see that everyone plays with at least one other child and that Cajus is the most friendly as he has a degree of 6. We see that some children only play with one other and that Marta and Basant are totally dependent on one another.

What does this tell us about the integration of the children in the group?

Can we make any conclusions about the formation of boy–girl friendships within the group, given only this information?

This method can be used by systems to connect people. If two people have several friends in common (mutual friends), they may also know one another and the system will recommend them to each other. Likewise, by analysing the number of connections or degrees of a node (person), we may be able to find the most connected, popular or influential person in that network.

PRACTICE EXERCISE

Make a network diagram to show the friendships in your class. Represent each class member as a vertex or node, and each friendship as a connecting line.

Sometimes, fans of TV shows create a network diagram to find out the most important character, so they can predict what might be coming next in the storyline and debate it online.

Try it with your own favourite novel or TV show. Who has the most connections to other characters? What might that mean for the plot? Does it make them the hero? Protagonist?

What about at your school? Who meets or sees the most people in one day? How could you judge this? Is it what you expected?

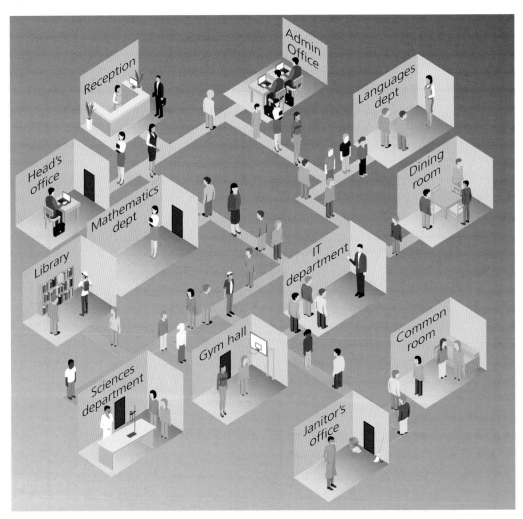

How can networks connect us in a global way?

MegNetz – connecting your customers all across the city

What is this image showing? Do you notice anything unusual about it? Who might be the 'best connected' in this network? Can you imagine this on a global scale?

Let's practise constructing and finding connections in networks.

Graph theory (study of all types of mathematical diagrams including networks) helps us to solve optimization problems – how to get the most or least of something from a set of given conditions. For example, a courier company might want to know how to deliver to every town in the area by covering the least distance, and a newly engaged couple may need to work out how to let everyone know they are getting married by making the fewest number of phone calls.

PRACTICE EXERCISE

1 Draw a network to represent your friendships, including different coloured or sized circles if needed.

2 Draw a network to show your family members.

3 A chess competition can be shown as the following network:

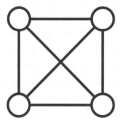

How many players were there? How many games were played?

4 For the following networks, count the number of different possible pathways to get from A to B. (You must always go from left to right, never backwards along an edge.)

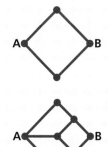

5 Redraw this map as a network.

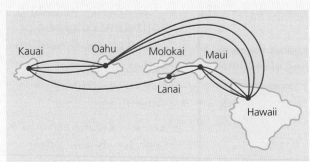

6 In some tournaments, the host nation gets an automatic qualification so they don't run the risk of getting knocked out in the first round. Identify which team is the host nation in this network.

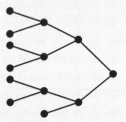

7 For the following networks, count the pathways to get from C to D. (You may go from top to bottom and right to left, but cannot travel on an edge that goes back or upwards.)

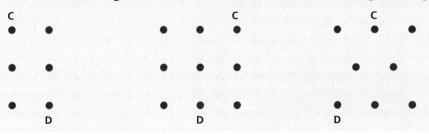

8 a A tennis league is staging a knock-out tournament for eight players. Draw a network to show how this might work, and state how many games would be played in total in this format.

b An alternative format for the tennis tournament is a 'round robin', in which everyone plays every other player to gain the most number of wins. Draw a network to show how this might work, and state how many games would be played in total in this format.

c Which network would you recommend and why?

9 The following networks show the possible outcomes for when a coin is tossed once, twice and three times. Fill in the number of paths to get to each outcome.

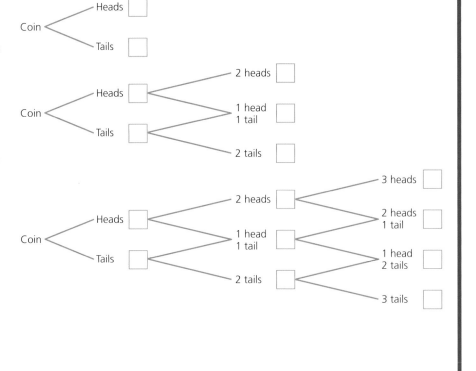

ACTIVITY: How do we decide what is best?

Let's consider a delivery person, Chas, who needs to deliver packages to several locations shown on this map. The distances between the towns (nodes) are also indicated.

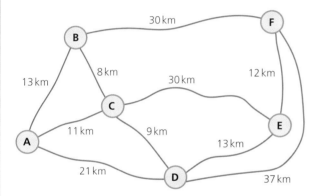

Part A: Delivering

Starting from the point A, how many paths can Chas take to get to town F?

Calculate the total distance for each route.

Which route is the shortest distance? Can you think of a reason why Chas might choose the shortest distance?

Which route is the longest distance? Can you think of a reason why Chas might choose the longest distance?

Again, starting at A, find a route that allows Chas to stop at **every** town. Calculate the distance for this route.

If Chas started at town F and wanted to drop off at every town, would the distance travelled be the same? Explain your answer.

Part B: Recalculating!

This diagram shows a subway network, and the number along each edge indicates the length of time it takes to travel along it in minutes (this is the **weighting**).

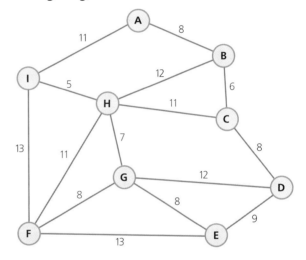

How many paths can you find to get from station I to station D?

What is the fastest way to get from station I to station D?

If station H closes, what is the fastest way to get from station I to station D?

Is it possible to visit every stop on the subway? If so, how long would it take? If not, why not?

What technology uses this same idea to recalculate routes?

Comment on the accuracy of your answers and whether they make sense in real life.

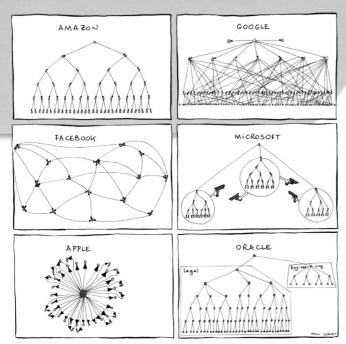

■ © Drawing by Manu Cornet

HISTORY'S HIDDEN NETWORKS

We have seen small, isolated networks of friends, towns and islands, but how do these networks work on a global scale? The historian Niall Ferguson suggests that we have paid far too much attention to states and individuals in history. He explains, in this video, that we might have paid too little notice to networks when trying to understand history and events: **www.youtube.com/watch?v=4cADSlk5CHU**

He uses networks to explain everything from great change, such as during the Enlightenment when ideas were passed between people through letters, to conspiracy theories like the one about The Illuminati. He creates network diagrams to suggest different ways of looking at events, such as one to prove why the religious Reformation was impossible to stop, despite mass executions. View this graph (slide 26) at **www.slideshare.net/niallcampbellferguson/the-square-and-the-tower-by-niall-ferguson**.

He also suggests that networks can help us understand very recent history. Look at the graph (top right) showing how messages spread and stay mostly inside the groups of people who share the same political outlooks.

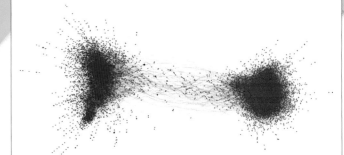

... which explains network polarization

Network graph of moral contagion shaded by political ideology. The graph represents a depiction of messages containing moral and emotional language, and their retweet activity, across all political topics (gun control, same-sex marriage, climate change). Nodes represent a user who sent a message, and edges (lines) represent a user retweeting another user. The two large communities were shaded based on the mean ideology of each respective community (blue = liberal, red = conservative mean).

What do you think of this way of approaching history? Evaluate the possible strengths and weaknesses of this approach.

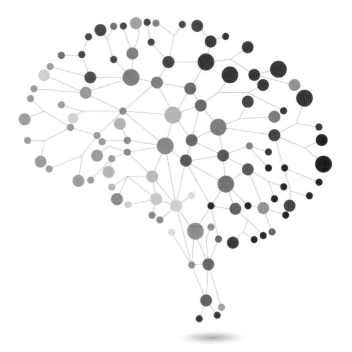

How does the brain operate like a network? How might the logic puzzles we tried better your mind on a larger scale?

ACTIVITY: Complete this network diagram

One of the most important ideas in the Middle Years Programme is the nature of interdisciplinary learning – that is, learning which takes place across and between different subjects. Draw a network diagram, placing each MYP subject at a vertex. Connect these vertices with edges. Along each edge, write one thing you have learned or observed that connects these subjects.

◆ Assessment opportunities

◆ In this activity you have practised skills that are assessed using Criterion A: Knowing and understanding.

ACTIVITY: Applying networks

Find an application of (use for) networks in each of these scenarios. Choose at least three scenarios and draw an example network for each of them:

- a delivery company planning for the week of Christmas deliveries
- a vice principal planning tutor groups (bearing in mind that certain students don't work well together)

- a team of police investigators trying to break up a group of criminals
- a boss deciding which team members to place together on three different projects
- a scriptwriter who just joined a show and needs to become quickly acquainted with the characters
- an advertising executive trying to decide where to place each advert in a series of commercials for a new product
- a basketball league organizer planning the schedule of games for the season.

◆ Assessment opportunities

◆ In this activity you have practised skills that are assessed using Criterion C: Communicating.

What are decision trees and how do they work?

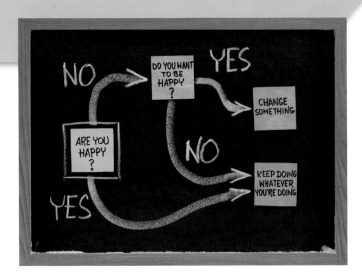

WHAT IS A DECISION?

Making a decision is the process of concluding or resolving an answer. A series of questions can help us to come to such an answer. Using a sequence of questions can be useful in problem-solving and logical thinking.

PRACTICE EXERCISE

1   Create a decision tree to classify types of angles (acute, right, obtuse and reflex).

2   Create a decision tree to classify types of quadrilaterals (square, rectangle, parallelogram, rhombus, kite and trapezium).

3   Create a decision tree to distinguish between a pyramid, sphere, prism, cylinder, cube and cuboid.

HOW DO DECISION TREES WORK?

A decision tree (or flowchart) is a series of questions that give different paths, depending on the answers given. You may be familiar with informal ones set out as quizzes, on social media for example. However,

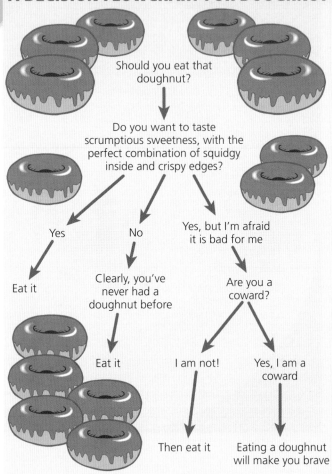

A DECISION FLOWCHART FOR DOUGHNUT

these are also useful mathematical tools that can be used to help solve complex problems or to categorize information quickly. The questions usually have simple yes or no answers, which make the alternative paths easy to define and follow.

For example, if we were to classify triangles using only questions with yes or no answers we could draw a tree like this:

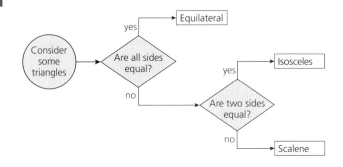

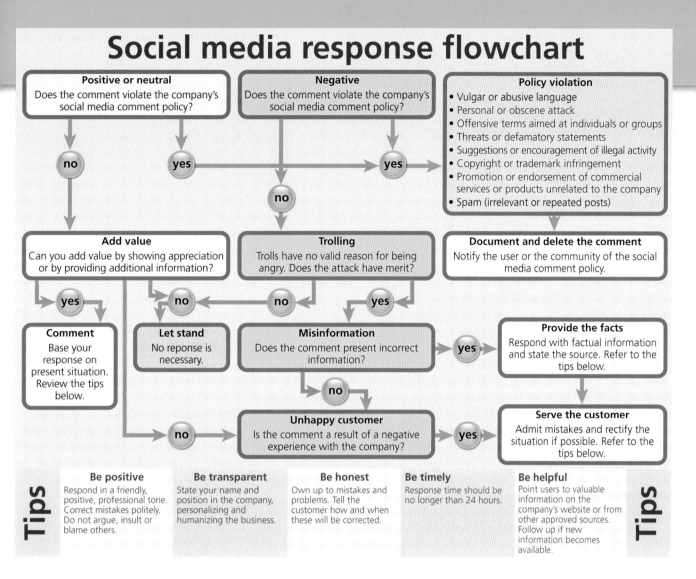

Social media response flowchart

Positive or neutral
Does the comment violate the company's social media comment policy?

Negative
Does the comment violate the company's social media comment policy?

Policy violation
- Vulgar or abusive language
- Personal or obscene attack
- Offensive terms aimed at individuals or groups
- Threats or defamatory statements
- Suggestions or encouragement of illegal activity
- Copyright or trademark infringement
- Promotion or endorsement of commercial services or products unrelated to the company
- Spam (irrelevant or repeated posts)

no   yes   yes   no

Add value
Can you add value by showing appreciation or by providing additional information?

Trolling
Trolls have no valid reason for being angry. Does the attack have merit?

Document and delete the comment
Notify the user or the community of the social media comment policy.

yes   no   no   yes

Comment
Base your response on present situation. Review the tips below.

Let stand
No reponse is necessary.

Misinformation
Does the comment present incorrect information?

yes

Provide the facts
Respond with factual information and state the source. Refer to the tips below.

no

Unhappy customer
Is the comment a result of a negative experience with the company?

no   yes

Serve the customer
Admit mistakes and rectify the situation if possible. Refer to the tips below.

Tips

Be positive
Respond in a friendly, positive, professional tone. Correct mistakes politely. Do not argue, insult or blame others.

Be transparent
State your name and position in the company, personalizing and humanizing the business.

Be honest
Own up to mistakes and problems. Tell the customer how and when these will be corrected.

Be timely
Response time should be no longer than 24 hours.

Be helpful
Point users to valuable information on the company's website or from other approved sources. Follow up if new information becomes available.

Tips

PRACTICE EXERCISE

Consider the following processes. Break each one down into a series of steps. Place them in order – which steps have to be done before the others? Can any of them be done at any time? Or at the same time as others?

Can you organize them as a network of activities or as an algorithm/flowchart?

1 Getting ready for school in the morning.
2 Downloading a new app on your phone.
3 Getting a new pet.
4 Planning a sleepover or party.
5 Doing a Service activity for a term.
6 Setting up a new account at an online shop (retailer).

ARE FLOW CHARTS ALSO ALGORITHMS?

Decision-making is a process – it's often helpful to think of the process as a series of steps. Some steps have to be done before others, some don't. Some steps can be **simultaneous**.

This is an algorithm being used by a company to help their customer service team work safely and efficiently online in the face of trolling, abuse or bullying.

How could you modify the algorithm to help students and young people stay safe online?

What is an algorithm and why should we care? Consider the information at this link:
www.khanacademy.org/computing/computer-science/ algorithms/intro-to-algorithms/v/what-are-algorithms

How can algorithms change things for us?

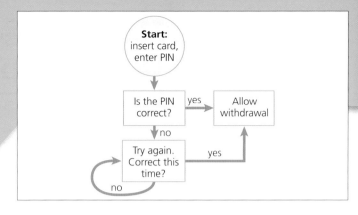

Algorithms can be defined as processes or rules to be followed by a human or computer to solve problems, make decisions or perform calculations.

Let's consider a simplified algorithm for entering a personal code (PIN) into a cash machine to withdraw money. Most machines allow three attempts in case the user makes a genuine mistake.

What problem can you see with the algorithm in the diagram at the top of this page? Can you suggest a solution?

How many outcomes are there in the pet-matching algorithm below? How many different paths are there to get to these outcomes?

Algorithmic thinking is a way of getting to a solution through the clear definition of the steps needed. They are instructions or rules that, if followed precisely (whether by a person or a computer), always lead to answers.

For example, we all learn algorithms for doing multiplication at school. If we (or a computer) follow the rules we were taught precisely, we can always get the correct answer to any multiplication problem. Think back to the various methods we used and compared for long multiplication in *Mathematics for the IB MYP 1*, Chapter 6. In that chapter, we saw four different algorithms (or step-by-step processes) to find the product of two large numbers.

Once we have mastered one of the algorithms, we don't have to work out how to do multiplication from scratch each time – it becomes part of our mathematical DNA and we can do it without too much thinking.

> *'The power of algorithmic thinking is that it allows solutions to be automated.'*
>
> Source: **https://teachinglondoncomputing.org/resources/developing-computational-thinking/algorithmic-thinking/**

But is that automation a good or a bad thing?

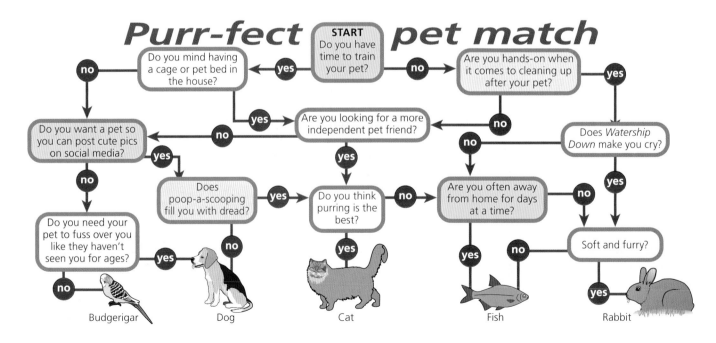

ACTIVITY: Flawed algorithms – a humorous approach

■ ATL

■ Critical-thinking skills: Recognize unstated assumptions and bias

Is the algorithm in the image at www.xkcd.com/1619/ (a decision tree to be used by a computer system) designed to assist or replace doctors?

Making random choices, write down two different paths through this algorithm. Were the outcomes different? How many possible outcomes are there? How many paths are there that could be followed?

The comic image is clearly not the real algorithm used by the IBM Diagnostic System which looks for errors, appropriately called 'Watson'. How do you know it is a joke? Give at least three examples of decision outcomes that make you believe the algorithm is only meant in a joking way.

What might the meaning be **behind** the humour? What does it tell you about the use of algorithms by machines compared to human decision-making?

◆ Assessment opportunities

◆ In this activity you have practised skills that are assessed using Criterion C: Communicating.

HOW ARE ALGORITHMS USED AROUND US?

Just like the activity opposite, dating websites use algorithms to predict good matches for people who are looking for a partner. These algorithms are often carefully guarded secrets. Other places we can see algorithms at work include:

■ when Netflix, Apple Music, Spotify or Amazon make recommendations for things you might like
■ when Google chooses the search terms that it thinks would best suit your search
■ when Snapchat finds your face
■ in other forms of facial recognition, such as at passport control
■ in 'pop-up' advertisements in your browser window that are supposed to be tailored to your interests
■ when airlines decide who to upgrade and who to 'bump off' a flight.

ACTIVITY: Mathematical speed dating

■ ATL

■ Creative-thinking skills: Make unexpected or unusual connections between objects and/or ideas

Some people find their perfect match through numbers. Here, you are not looking for romance, but for your number buddy!

Set up the class with lots of individual tables, two seats at each. Half the class members stay seated, and the other half move when the teacher rings a bell after one minute. For every 'number date' you must ask the following questions and write down your partner's response. They will ask you the same questions.

● **What is your favourite number? Why?**
● **Do you have a lucky number? What is it?**
● **Which times table do you know the best?**
● **What is the number of your birthday month?**

When each of the 'movers' has met every 'sitter', you should all come together to compare results.

Did you find your number buddy, a perfect match? Did anyone find a perfect numerical match? How close were your answers to the others' answers? How could you calculate the closest person to you?

◆ Assessment opportunities

◆ In this activity you have practised skills that are assessed using Criterion A: Knowing and understanding.

What do we mean by 'invisible algorithms'?

THINK-PAIR-SHARE

People who want to be fair in their decision-making try to remove emotion from the process. This **objective** approach aims to remove **bias**. Do you think this is possible? Is this the purpose of automating (or 'computizing') decision-making?

ANALYSIS FOR DECISION-MAKING

ACTIVITY: True or false?

ATL

■ Critical-thinking skills: Recognize unstated assumptions and bias

'I don't think mathematical models are inherently evil – I think it's the ways they're used that are evil,' says mathematician Cathy O'Neil, author of the book *Weapons of Math Destruction*. In her book she details how algorithms are all around us, making decisions with results that humans have to carry out. She uses the idea that these algorithms are widespread, mysterious and destructive to explain the effect they might have on our lives. Listen to this podcast and consider her arguments before completing these activities.

Source: https://99percentinvisible.org/episode/the-age-of-the-algorithm/

Which of the following statements are true, according to the podcast mentioned here?

● **Using an algorithm to make decisions means they are free from human bias or error.**
● **An algorithm is not a step-by-step guide to follow to make a decision.**
● **Racist courtroom judges no longer exist because of algorithms.**
● **An algorithm is used to decide what college you will get into.**
● **Algorithms are more trouble than they are worth.**
● **Nothing is neutral in algorithms.**
● **We know who the algorithms are failing and who is winning or losing because of them.**
● **People are often hired, or not, based on a personality test algorithm.**
● **People do not accept the results of algorithms and question them because they feel so uncomfortable with mathematical ideas.**

◆ Assessment opportunities

◆ In this activity you have practised skills that are assessed using Criterion D: Applying mathematics in real-life contexts.

SUMMATIVE ASSESSMENT

Use these problems to apply and extend your learning in this chapter. The problems are designed so that you can evaluate your learning at different levels of achievement in Criterion C: Communicating and Criterion D: Applying mathematics in real-life contexts.

In the following assessments, we will consider the personal, the local and the global aspects of networks and algorithms.

Part A: The personal

Which Learner Profile are you?

The following image shows the 10 different Learner Profile attributes. You will use this to create two different products, one network and one algorithm, using your knowledge from this chapter.

1 **Network**: Your teacher will provide a large version of this image. Connect two Learner Profile attributes with a string or ribbon, attaching the name of a class member who demonstrates them both. Repeat this process for each person in the class. Include your teacher as well.

2 **Algorithm** – There are many popular quizzes on the internet that claim to help you find out which 'type' you are – for example, 'Where should you live?', 'Which character from *Stranger Things* are you?'

Your task is to design a quiz that will help someone decide their Learner Profile. The 'Which Learner Profile are you?' quiz should use a decision tree or an algorithm with paths that end at each of the Learner Profile characteristics, determined by questions with 'yes' or 'no' answers only.

➤

Part B: The local

Your town is bidding to host a very important event and the judges must visit to decide if it is a suitable location. As the event coordinator, you must plan a route to show the judges the best of your city.

Your task is to design, and represent, excellent routes for the following different scenarios:

1 a large sports tournament, like a sporting World Cup or Olympics
2 a cultural event (for example, City of Culture)
3 a 'Tidy Towns' competition
4 a food festival, so the judges want to visit the top five recommended restaurants in your town or city
5 a pedestrian-friendly city, so the judges wish to travel by foot only and on car-free roads if possible.

Part C: Global networks

Consider again the Statement of Inquiry for this chapter:

Global networks are built on logic and are changing the way we handle data, make decisions and design models.

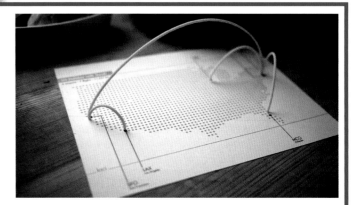

To complete this part of the assessment you should research and investigate how connected your location is with others, physically and geographically.

Construct network diagrams to represent how your home town or city is connected to others. Show:

1 all other towns and cities that it is connected to by road (or by another form of transport) by **one degree** – that is, those locations with a **direct** connection
2 all other towns and cities that it is connected to by **two degrees** – that is, connected **via** another town or city
3 how far you can get from your home location in one flight
4 how far you can get from your home location in two flights.

You will need a world map for tasks 3 and 4. Use the illustrations above and below for inspiration.

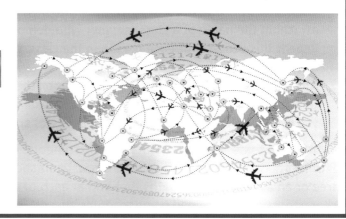

Reflection

Use this table to reflect on your own learning in this chapter.					
Questions we asked	Answers we found	Any further questions now?			
Factual: What are networks? How does logic lead to graph theory? What are decision trees and how do they work? What do we mean by 'invisible algorithms'?					
Conceptual: How can games be based on logic? Can puzzles be modelled? How can algorithms change things for us?					
Debatable: How old are games? How can networks connect us in a global way? Are they always positive?					
Approaches to Learning you used in this chapter:	Description – what new skills did you learn?	How well did you master the skills?			
		Novice	Learner	Practitioner	Expert
Critical-thinking skills					
Media literacy skills					
Creative-thinking skills					
Affective skills					
Transfer skills					
Learner Profile attribute(s)	Reflect on the importance of being principled and balanced for your learning in this chapter.				
Principled					
Balanced					

4 What are the chances?

○ **Patterns** found in **relationships** can be **generalized** to help us make predictions for **personal** gain.

CONSIDER THESE QUESTIONS:

Factual: How do we know what to expect?

Conceptual: Are all probabilities created equal? Why does mathematics sometimes overcomplicate simple problems? How does grouping make probability easier? How can trees serve as metaphors?

Debatable: Is probability just for fun? Does all probability have to be theoretical?

Now **share and compare** your thoughts and ideas with your partner, or with the whole class.

○ IN THIS CHAPTER, WE WILL ...

- **Find out** how to calculate simple probabilities.
- **Explore** different ways to visualize outcomes.
- **Take action** by putting risk into perspective for those who don't take the time to look into it.

These Approaches to Learning (ATL) skills will be useful …

- Communication skills
- Information literacy skills
- Creative-thinking skills
- Critical-thinking skills

We will reflect on these Learner Profile attributes …

- **Risk-taker** – We approach uncertainty with forethought and determination; we work independently and co-operatively to explore new ideas and innovative strategies. We are resourceful and resilient in the face of challenges and change.
- **Caring** – We show empathy, compassion and respect. We have a commitment to service, and we act to make a positive difference in the lives of others and in the world around us.

Assessment opportunities in this chapter:

- **Criterion A:** Knowing and understanding
- **Criterion B:** Investigating patterns
- **Criterion C:** Communicating
- **Criterion D:** Applying mathematics in real-life contexts

THINK-PAIR-SHARE

Have you ever asked yourself: 'What are the chances that [something will happen]?' List as many variations of that question as you can. Here are a few to start you off:

What are the chances that …

- … I will run into [name of my friend/crush] in the hall between classes?
- … the bus will be on time today, as I am running late?
- … the guess I made on my multiple choice test this morning was right?
- … my best friend and I will wear the same thing to the next school dance?

With a partner, think of some more scenarios.

Can you attach a numerical value to each of these questions? If so, estimate the values, and then discuss as a class!

PRIOR KNOWLEDGE

Reflect on what you already know about:

- comparing, adding, subtracting and multiplying fractions, decimals and percentages
- applying the order of operations
- set theory.

KEY WORDS

likelihood	dice	die

Is probability just for fun?

Probability is the likelihood of an event occurring. The event can be anything at all, from your dog eating your homework to a tsunami occurring or your favourite team winning the championship. Or it could be something simpler – like rolling a number 4 on a die or flipping 'heads' on a coin. It is a common misconception that probability is only relevant to gamblers in casinos. Though gambling offers many scenarios in which to visualize aspects of probability theory, there are many more situations to which it applies.

■ Which day is least likely to have rain?

■ What is the likelihood of having all four aces in your hand from a 52-card deck?

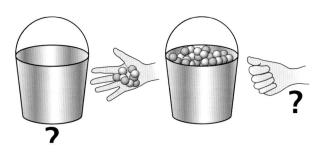

Statistics
Given what is in your hand, can you work out what is in the bucket?

Probability
Given what is in the bucket, can you work out what is in your hand?

■ The difference between probability and statistics

We have studied statistics in both *Mathematics for the IB MYP 1* and *2*, and the two topics of statistics and probability are often grouped together into one strand. Both involve the study of the relative frequency of events, but statistics tends to look backwards, analysing data in order to make informed future predictions, while probability looks forward to measure the likelihood of events primarily based on an assumption of randomness. Both have important uses but knowing which to use depends on the sort of problem we are trying to solve and what information is available to us.

■ What is the probability of pulling a muscle while in the sidecar pose?

■ What errors did each character here make about probability?

THINK-PAIR-SHARE

Look at the phone app, playing card and yoga images again. This time, for each, state whether you think the probabilities are high, low, even, certain or impossible.

Q. What event has a probability of 1?

A. *The sun will come up in the morning.* ✓

Q. What event has a probability of 0?

A. *You will win teacher of the year.* ✗

A. *You will be nominated student of the month.*

■ How would you explain this cartoon to somebody who doesn't know about probability?

Some things are certain, some are impossible, but most are somewhere in between. This 'in between' is assigned a number between 0 and 1, or between 0% and 100%. A 0% or 0 probability indicates an impossible event, and a 100% or 1 indicates certainty. What sorts of values are there in between? Can you label where 'probably', 'unlikely' and 'an even chance' might be on the probability scale below?

Assign a number between 0 and +1 which indicates your estimated probability of:
- seeing the dentist this year
- going ice skating this month
- eating four servings of vegetables tomorrow
- a thunderstorm in the Sahara desert
- an airplane landing in front of your house
- losing power this month
- the Sun rising tomorrow.

ACTIVTY: Putting things into perspective

■ **ATL**

■ Communication skills: Understand and use mathematical notation

Draw a probability scale like the one below and label the probabilities of the listed events on your scale. It is similar to a timeline! The first one has been done for you. Think of a solution if there is not enough room to write out each event in full.

- **Your online order will arrive on time.**
- **A helicopter will crash onto your house.**
- **You will be late to at least one class this week.**
- **You will speak to someone today.**
- **A dinosaur will attend class this week.**
- **This evening's news will cover a happy story.**

Now add three of your own events, each with a different likelihood.

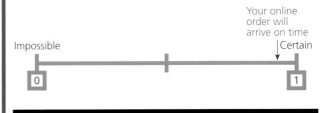

◆ **Assessment opportunities**

◆ In this activity you have practised skills that are assessed using Criterion B: Investigating patterns and Criterion C: Communicating.

Why does mathematics sometimes overcomplicate simple problems?

To get used to some terminology, let's look at one of the simplest examples of probability: the coin toss. The **event** of tossing a coin has two possible **outcomes** – we could get a head, or we could get a tail. Suppose our **desired outcome** – the outcome whose probability we want to determine – is to toss a head. Let's show this in a **sample space** (a space or box that contains all our possible outcomes).

Head   Tail

We are trying to 'find the probability of tossing a head'. This can be written simply as P(H), where H represents the outcome of tossing a head. Let's look closely at that notation.

Tossing a head

Probability

$$P(H)$$

of

To calculate any theoretical probability, no matter how complex, we think about a calculation like this:

$$\frac{\text{\# of possible desired outcomes (that is \# of ways to get what we are looking for)}}{\text{total \# of possible outcomes (total \# of things that can happen)}}$$

We can list the numbers we need in the coin-tossing problem:

- One desired outcome: H
- Two possible outcomes: H, T

$$P(H) = \tfrac{1}{2} \text{ or } 0.5 \text{ or } 50\%$$

ⓘ Note that the 'desired outcome', is not always a desirable one! If seeking the probability of contracting a disease, for example, contracting it is still regarded as the 'desired outcome'.

Examples

Cover up the solutions and try these examples on your own first!

Draw the sample space for each scenario, circle the desired outcome and give the probability (in simplest form):

1   Guessing False correctly in a True/False quiz

2   Rolling a 2 on a 6-sided die

3   Rolling an even number on a 6-sided die

4   Rolling a 5 or higher on a 6-sided die

5   Randomly selecting a prime number from the numbers 1–20

6   Selecting one of the five winning numbers in a lottery that gives 49 choices, whose winning numbers are 4, 12, 19, 32, 35.

Solutions

1

There is one way to get our desired outcome (false), and there are two outcomes in the sample space, so $P(F) = \frac{1}{2}$

2

There is one out of six possible outcomes, so $P(2) = \frac{1}{6}$

3

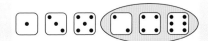

There are three out of six possible outcomes, so $P(\text{even}) = \frac{3}{6} = \frac{1}{2}$

4

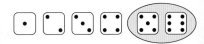

There are two out of six possible outcomes, so $P(5, 6) = \frac{2}{6} = \frac{1}{3}$

5

$$9 \quad 14 \quad 7 \quad 19 \quad 1 \quad 20$$
$$12 \quad \quad 2 \quad \quad \quad 8$$
$$4 \quad 6 \quad 11 \quad 17$$
$$3$$
$$15 \quad 18 \quad 13 \quad 5 \quad 16 \quad 10$$

There are eight out of 20 possible outcomes, so $P(\text{prime}) = \frac{8}{20} = \frac{2}{5}$

6

$$15 \quad 6 \quad 18 \quad 11 \quad 13 \quad 16 \quad 21$$
$$34 \quad 14 \quad 7 \quad 29 \quad 8 \quad 3 \quad 40$$
$$42 \quad 49 \quad 4 \quad 12 \quad 20 \quad 23 \quad 17$$
$$28 \quad 37 \quad 19 \quad 32 \quad 5 \quad 30 \quad 24$$
$$33 \quad \quad 35 \quad 38$$
$$46 \quad 41 \quad 31 \quad 39 \quad 45 \quad 26 \quad 25$$
$$36 \quad 44 \quad 43 \quad 47 \quad 22 \quad 10 \quad 48$$
$$9 \quad 27 \quad 1 \quad \quad \quad 2$$

There are five out of 49 possible outcomes, so $P(\text{selecting one winning number}) = \frac{5}{49}$

PRACTICE EXERCISE

Die 1

	1	2	3	4	5	6
1	2	3	4	5	6	7
2	3	4	5	6	7	8
3	4	5	6	7	8	9
4	5	6	7	8	9	10
5	6	7	8	9	10	11
6	7	8	9	10	11	12

Die 2

■ Sample space for the sum of two dice

For the event of rolling two dice, answer these questions.

1   How many elements are in the sample space when considering the sums of two dice?

2   Find the probability of rolling 'snake eyes' (two ones).

3   Find the probability of rolling 'doubles' (two of the same number).

4   Find the probability of rolling a sum that is less than 5.

5   What is the sum with the highest probability?

ACTIVITY: Are they 'the One'?

The 'Sheng Xiao' – known as the Chinese Zodiac – comprises 12 animals that represent each of 12 years in the Chinese lunar calendar. Which animal are you?

鼠
Rat
1912, 1924, 1936, 1948, 1960, 1972, 1984, 1996, 2008

牛
Ox
1913, 1925, 1937, 1949, 1961, 1973, 1985, 1997, 2009

虎
Tiger
1914, 1926, 1938, 1950, 1962, 1974, 1986, 1998, 2010

兔
Rabbit
1915, 1927, 1939, 1951, 1963, 1975, 1987, 1999, 2011

龍
Dragon
1916, 1928, 1940, 1952, 1964, 1976, 1988, 2000, 2012

蛇
Snake
1917, 1929, 1941, 1953, 1965, 1977, 1989, 2001, 2013

馬
Horse
1918, 1930, 1942, 1954, 1966, 1978, 1990, 2002, 2014

羊
Goat
1919, 1931, 1943, 1955, 1967, 1979, 1991, 2003, 2015

猴
Monkey
1920, 1932, 1944, 1956, 1968, 1980, 1992, 2004, 2016

雞
Rooster
1921, 1933, 1945, 1957, 1969, 1981, 1993, 2005, 2017

狗
Dog
1922, 1934, 1946, 1958, 1970, 1982, 1994, 2006, 2018

豬
Pig
1923, 1935, 1947, 1959, 1971, 1983, 1995, 2007, 2019

People who are born in a given year are said to share the characteristics of that year's animal. In ancient times, and sometimes still today, people would often refer to the Chinese Zodiac compatibility chart before beginning a romantic relationship.

Are you two compatible?

	Rat	Ox	Tiger	Rabbit	Dragon	Snake	Horse	Goat	Monkey	Rooster	Dog	Pig
Rat	Average	Very compatible	Average	Very compatible	Very compatible	Good friend	Not compatible	Enemies	Complementary	Not compatible	Complementary	Average
Ox	Very compatible	Average	Not compatible	Complementary	Not compatible	Good friend	Not compatible	Not compatible	Very compatible	Very compatible	Average	Enemies
Tiger	Average	Not compatible	Average	Average	Not compatible	Not compatible	Very compatible	Good friend	Not compatible	Average	Average	Average
Rabbit	Very compatible	Complementary	Average	Average	Average	Average	Average	Very compatible	Average	Not compatible	Very compatible	Very compatible
Dragon	Very compatible	Not compatible	Average	Average	Good friend	Very compatible	Not compatible	Average	Complementary	Good friend	Not compatible	Good friend
Snake	Good friend	Complementary	Not compatible	Average	Very compatible	Average	Good friend	Very compatible	Not compatible	Complementary	Not compatible	Not compatible
Horse	Not compatible	Not compatible	Very compatible	Average	Not compatible	Good friend	Average	Complementary	Average	Average	Average	Complementary
Goat	Enemies	Not compatible	Good friend	Very compatible	Average	Very compatible	Complementary	Average	Average	Average	Not compatible	Very compatible
Monkey	Complementary	Very compatible	Not compatible	Complementary	Complementary	Good friend	Average	Average	Average	Good friend	Complementary	Average
Rooster	Not compatible	Very compatible	Average	Not compatible	Complementary	Very compatible	Not compatible	Average	Average	Average	Not compatible	Average
Dog	Complementary	Average	Average	Very compatible	Not compatible	Not compatible	Average	Not compatible	Complementary	Not compatible	Average	Average
Pig	Complementary	Enemies	Very compatible	Very compatible	Good friend	Not compatible	Complementary	Very compatible	Not compatible	Average	Average	Good friend

Key

♥ Very compatible ☺ Complementary 👍 Good friend

😐 Average 👎 Not compatible ♥☠ Possibly a perfect match, most likely enemies!

1 What types of relationships are in the sample space?

2 If randomly selected, what is the probability that a snake will be matched up with a good friend?

3 What is the probability that two randomly selected people will be 'very compatible'?

4 Which sign has the highest probability of meeting someone who is
   - 'very compatible'?
   - 'not compatible'?

5 What big mathematical assumption are we making when answering the previous three questions?

Are all probabilities created equal?

▼ Links to: Science

Autosomal recessive traits are physical traits or diseases that are passed down through families. This is dealt with more extensively in *Sciences for the IB MYP 2: by Concept, Chapter 3*. The infographic below shows the likelihood of different genetic outcomes.

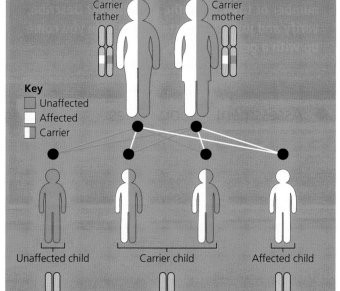

Carrier father   Carrier mother

Key
- ☐ Unaffected
- ☐ Affected
- ☐ Carrier

Unaffected child   Carrier child   Affected child

25% probability   50% probability   25% probability

■ How is probability useful in determining whether a child will carry any of these traits?

INDEPENDENT VERSUS DEPENDENT EVENTS

If you were to roll a fair die three times and get three fives in a row, what would the probability be of getting another five? Would the past three flips affect the next flip?

No!

Each flip is completely random and unrelated to previous outcomes. In other words, each flip is **independent** of other flips.

■ Independent event

■ Dependent event

THINK-PAIR-SHARE

Think about the scenarios below. With a partner, discuss whether each set of events is dependent or independent:

- **rolling a 5 on a die and flipping a head on a coin**
- **drawing a king from a deck of cards then, without putting it back, selecting another king**
- **rain on the day and the parade being cancelled**
- **earning a 5 on an English assessment and earning a 7 on a Mathematics assessment**
- **sunny on a certain day and getting a job offer**
- **randomly selecting pink socks and brown shoes.**

Now come up with some of your own examples of independent and dependent events.

On the other hand, if the event was tossing a basketball into a basket from the same spot, it may be easier the third time than it was the first because you have a better feel for it. Each throw depends on how the previous throw went, and you can make adjustments based on what happened. This would be considered a **dependent** event.

ACTIVITY: Examining multiple independent events

■ ATL

■ Creative-thinking skills: Use brainstorming and mind mapping to generate new ideas and inquiries; Create novel solutions to complex problems

1 Determine the probability of tossing a head by:
   a drawing the sample space and circling the desired outcome.
   b calculating
   $$\frac{\text{desired outcome}}{\text{total number of outcomes}}$$

2 Determine the probability of tossing **two** heads in a row by:
   a listing all possible outcomes ((H, T), (H, H) and so on) and circling the desired outcome.
   b calculating
   $$\frac{\text{desired outcome}}{\text{total number of outcomes}}$$

3 Determine the probability of tossing **three** heads in a row by:
   a listing all possible outcomes ((H, T, T), (H, H, H) and so on) and circling the desired outcome.

   b calculating
   $$\frac{\text{desired outcome}}{\text{total number of outcomes}}$$

4 Copy and complete the table below:

Number of tosses	P(H)
1	
2	
3	
4	
n	

5 Have you noticed a relationship between the number of tosses and the probability? Describe, verify and justify your statement. Can you come up with a general equation?

◆ Assessment opportunities

◆ In this activity you have practised skills that can be assessed using Criterion B: Investigating patterns.

Examples

Ryan agrees to let his four-year-old daughter pick out his clothes for the day. Her options are pictured here.

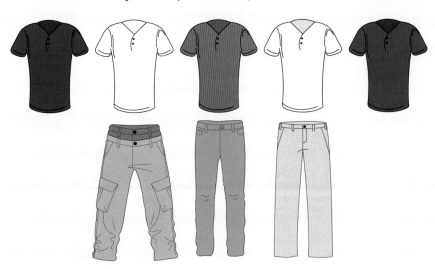

1 Determine the number of elements in the sample space, n(S).

2 What is the probability his daughter will select the one outfit Ryan doesn't want – purple shirt and grey cargo pants?

3 What is the probability she will select Ryan's favourite combination – a blue shirt and khakis?

4 What is the probability she will select either his favourite combination **or** the outfit he doesn't want?

5 What is the probability the outfit will include a pair of jeans?

6 What is the probability she will **not** select his favourite outfit?

Solutions

1 To determine n(S), we can list all possible combinations:

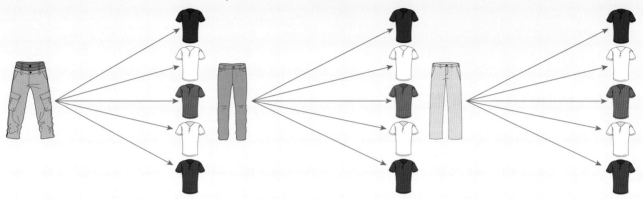

If we count all these combinations (or count the arrows), we'll see that we have a total of 15 possible outfit selections. This was a bit time-consuming and mentally easy to do – but what if Ryan had 20 different kinds of pants and 45 different shirts? It could take days to list out the outfit combinations. Is there a faster way? Have you noticed a relationship between the number of pants, the number of shirts and the final number of outfits?

There are three pairs of pants and for each pair there are five shirts to match, making 15 combinations:

$3 \times 5 = 15$

In future, let's save ourselves the trouble and simply multiply!

2 There is only one purple shirt and one pair of grey cargo pants in Ryan's closet, so there is only one way to achieve the desired outcome. Thus, P(purple and cargos) = $\frac{1}{15}$. Another way to look at this is to take the probability of selecting the purple shirt and multiply it by the probability of selecting the cargos:

$\frac{1}{5} \times \frac{1}{3} = \frac{1}{15}$

3 As there are two ways of selecting a blue shirt (light and dark), P(blue) = $\frac{2}{5}$

So, P(blue shirt) × P(khakis)

$= \frac{2}{5} \times \frac{1}{3}$

$= \frac{2}{15}$

4 Since there are three desired outcomes – purple shirt and cargos, light blue shirt and khakis, dark blue shirt and khakis – the probability of selecting one of these is $\frac{3}{15}$ or $\frac{1}{5}$. Another way to look at this is if we add our results from question 2 and question 3, we get

$\frac{1}{15} + \frac{2}{15} = \frac{3}{15}$ or $\frac{1}{5}$

5 As the question does not specify a shirt, we can choose any one of five shirts to go with the cargos, so P(cargos) = $\frac{5}{15}$ or $\frac{1}{3}$.

6 We **could** go through the probabilities of every combination but the favourite outfit and add them all together, but when a problem involves NOT selecting something, it is often simpler to calculate the probability of selecting what we don't want, and subtracting this from 1. Why? Because the probabilities of all possible outcomes must always add to 100%, or 1.

We know from question 3 that the probability of selecting Ryan's favourite outfit is $\frac{2}{15}$. Thus, the probability of the **opposite** occurring, of selecting his favourite outfit, is

$1 - \left(\frac{2}{15}\right) = \frac{13}{15}$

FUNDAMENTAL COUNTING PRINCIPLE

The previous example shows that there are shortcut ways of determining the number of outcomes and probabilities to avoid time-consuming counting of lists or parts of diagrams. Essentially we can summarize these short cuts as follows.

AND → Multiply

When determining the probability of two or more independent events **all** occurring, often phrased as 'the probability of x **and** y **and** z', we multiply those probabilities together. For instance, the probability of flipping two heads in a row is

$$\frac{1}{2} \times \frac{1}{2} = \frac{1}{4}$$

Flipping 10 heads in a row is

$$\frac{1}{2^{10}} = \frac{1}{1024}$$

This is often referred to as the **fundamental counting principle**.

OR → Add

When determining the probability of one independent event **or** another occurring, we add their probabilities. For instance, the probability of rolling a four **or** an odd number is

$$\frac{1}{6} + \frac{3}{6} = \frac{4}{6} \text{ or } \frac{2}{3}$$

Complement

The **complement** of an event is simply the opposite. The complement of rolling a four is rolling a one, two, three, five or six. The probability of an event and the probability of its complement always add to 1, as together they encompass all possible outcomes.

NOT → Subtract from one

The probability of an event's complement occurring can be determined simply by subtracting the probability of the event from 1. The probability of **not** rolling a four is

$$1 - \frac{1}{6} = \frac{5}{6}$$

In some cases, such as the probability of **not** flipping a head, this technique is redundant as it takes more time to subtract $1 - \frac{1}{2}$ than to calculate the probability of tails, $\frac{1}{2}$.

A popular English Christmas carol (song) tells the story of unusual and lavish gifts that are sent to a significant other on each of the 12 days of Christmas, which for Christians start at Christmas day (the birth of Christ) and end with Epiphany (the coming of the Magi).

William wishes to re-enact the events of the song, and purchases all of the gifts in advance. Stored in his garage, as well as food and lodging for all the live gifts, are:

- 1 partridge in a pear tree
- 2 turtle doves
- 3 French hens
- 4 calling birds
- 5 gold rings
- 6 geese laying
- 7 swans swimming
- 8 maids milking
- 9 ladies dancing
- 10 lords leaping
- 11 pipers piping
- 12 drummers drumming.

The TWELVE DAYS of CHRISTMAS

What is the probability, if three gifts are selected at random, that they are all birds of some kind?

Before proceeding with this question, we must ask, were the gifts selected:

- one at a time and each returned before selecting the next
- one after another without putting anything back
- at exactly the same time?

DISCUSS

Which scenario offers a greater probability, replacement or no replacement? Why do you think this is?

PRACTICE EXERCISE

1 Determine the number of possible outcomes when flipping a coin and rolling a die.

2 What is the complement of rolling an even number?

3 Calculate the probability of rolling a 2 and selecting a blue marble from the bag pictured.

4 Calculate the probability of rolling a 6 or selecting a yellow marble.

5 Calculate the probability of not selecting a yellow marble.

6 Calculate the probability of **not** rolling a 2 and **not** selecting a red marble.

7 Calculate the probability of rolling an even number and selecting a blue or red marble.

8 Calculate the probability of rolling a sum of at least 8 with two six-sided dice.

9 Come up with your own 'and', 'or' and 'complements' question. Provide solutions. Trade questions with a partner to check each other's solutions. Finally, grade your partner on Criterion A. Submit your blank question and solution, as well as your partner's marked paper.

With replacement	Without replacement
– one gift selected, noted, and returned to the garage before picking the next	– three gifts selected either all at once or without returning any to the garage
n(S) = 1 + 2 + 3 + 4 + 5 + 6 + 7 + 8 + 9 + 10 + 11 + 12	n(S) = 1 + 2 + 3 + 4 + 5 + 6 + 7 + 8 + 9 + 10 + 11 + 12
= 78	= 78
n(B) = number of birds: 1 + 2 + 3 + 4 + 6 + 7 = 23	n(B) = number of birds after each selection: ■ First there are 23 birds and 78 gifts ■ If the first gift selected is a bird, then there are 22 birds left in the garage and 77 gifts ■ If the second gift selected is **also** a bird there are 21 birds and 76 gifts remaining
P(B then B then B) = $\frac{23}{78} \times \frac{23}{78} \times \frac{23}{78}$ = $\frac{12\,167}{474\,552}$	P(B and B and B) = $\frac{23}{78} \times \frac{22}{77} \times \frac{21}{76}$ = $\frac{10\,626}{456\,456} = \frac{23}{988}$

Does all probability have to be theoretical?

In situations that are more complicated, which involve many possible and less predictable factors such as how humans react, theoretical probability simply isn't enough. We can instead turn to data and past experience to determine the **experimental probability**. Some data may be intuitive – the likelihood of running into heavy traffic between 5 p.m. and 6 p.m. is high, based on past experiences of driving at this time – and some data are collected in controlled environments, such as the study of the effects of a new medicine.

Experimental data from the American Cancer Society suggest both men and women have a much higher probability of contracting lung cancer than oesophageal cancer. How is this knowledge helpful to medical practitioners?

ⓘ Simulations

In many cases, determining the probability of an event by experimentation is too complicated or too risky. We know, for instance, that when guessing on a True–False quiz we have a 50% chance of guessing correctly and 50% of guessing incorrectly. If we do this 100 times to determine the experimental probability, we may risk a failing grade! Or at least gaining one that is not representative of our true knowledge. We can, instead, use a **model** with the same equal probabilities as our True–False quiz to **simulate** the outcomes. Flipping a coin has the same 50% probability of each outcome – we could say 'heads' represents a correct answer and 'tails' represents an incorrect answer. You could flip the coin 100 times to give an experimental probability without actually guessing on the quizzes!

It is extremely difficult for theoretical probability to replace experimental probability, but can experimental probability be applied to simple things like flipping a coin or rolling a die? Sure it can! However, sometimes a simulation is useful and other times it is only circumstantial.

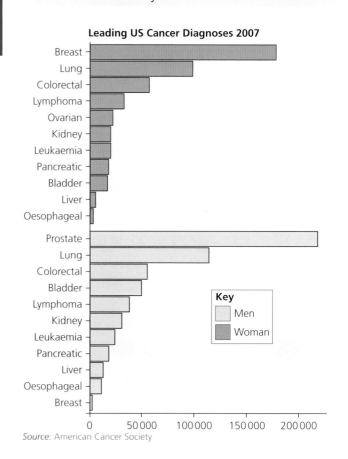

Leading US Cancer Diagnoses 2007

Source: American Cancer Society

ACTIVITY: Simulating a coin toss

A long time ago, textbooks used to ask students to flip coins hundreds of times to get a point across. Luckily for you, we now have technology to do that for us.

Select your preferred mathematical software – this may be a graphing calculator, a spreadsheet program, Desmos or some other similar program – and determine how to generate a random number. This means the program will select any number and display it, and there is no way for you to predict what that number could be.

Determine how to set boundaries for that random number, so it can only be 1 or 2. In most cases we cannot ask the computer to randomly select heads or tails (graphing calculators do, but we want to generalize across many contexts – not just coin flipping), so we will use 1 to represent heads, and 2 to represent tails.

Let's determine the experimental probability of flipping a head. 'Perform' three flips – in other words, simulate them by randomly generating three numbers. How many heads came up in the three trials? How does that number compare to the theoretical probability of flipping heads?

Repeat with 10 flips. Then 100, 1000, 10000. What do you notice?

How might you use random number generation to simulate rolling a die? Drawing a queen from a deck of cards? A male or female person winning the lottery?

WHAT MAKES YOU SAY THAT?

Adina participates in Hallowe'en – a day when children knock on doors in costume and are given a small treat. Adina is expecting lots of children to call at her home so she starts with 50 orange-wrapped chocolate bars and 50 red-wrapped chocolate bars. She calculates the probability of selecting an orange bar:

$\frac{50}{100} = \frac{1}{2}$ or 50%

No kids have shown up at her house and she is bored, so to pass the time, she tries an experiment. She randomly selects a chocolate bar, writes her result and places the bar back in the bucket. She does this 10 times and she ends up having selected orange bars eight times and red twice. This suggests an 80% probability that she would select an orange bar.

Think about the theoretical versus experimental results: which is more reliable after only 10 trials?

The night grows long. Adina repeats her experiment 1000 times and pulls out 860 orange bars. This suggests an 86% chance she will select an orange bar.

Again, consider theoretical versus experimental: which is more accurate? What are some possible reasons for this wide discrepancy?

■ What does this meme say about experimental versus theoretical probability?

ACTIVITY: Experimental versus theoretical probability

ATL

- Information literacy skills: Collect, record and verify data; Make connections between various sources of information; Process data and report results

Come up with your own simple event to analyse. It must be something that can be carried out either experimentally or as a simulation. Calculate the experimental and theoretical probabilities, and determine which – in your case – is the more reliable. Justify your response.

For example, is the spinner on the TV programme *Wheel of Fortune* fair? Calculate the theoretical probability of spinning a positive result after looking at an online image of the spinner. Then watch an episode and record the number of positive versus negative results to calculate an experimental probability.

◆ Assessment opportunities

- ◆ In this activity you have practised skills that are assessed using Criterion C: Communicating and Criterion D: Applying mathematics in real-life contexts.

MEET A MATHEMATICIAN: JEAN-BAPTISTE MICHEL

Learner Profile: Caring

How can you chart cultural and societal changes? When did French and German words creep into the English language? Why? When did the people of the USA first consider themselves to belong to one unified country? How can you determine these answers? Why is it important?

Jean-Baptiste Michel came up with a way to study cultural and societal trends using big data. He has coined the term 'culturomics' to describe the quantification (giving numbers to) of culture, and does it by analysing 12% of all books ever printed.

In his talk *Bigger Data, Better World* (www.thelavinagency.com/speakers/jeanbaptiste-michel#39123987140533190) he discusses his partnership with Google to create the Ngram Viewer, which graphs the appearance of selected word(s) across the centuries. This can give a good picture of society's concerns at various points in time. How does this graph show that the words used imply a change of identity for a whole nation? Michel follows up with such examples in his book, *Uncharted: Big data as a lens on human culture*, and created art based on his findings that is displayed at the Whitney Museum in New York: I wish I could be exactly what you're looking for.

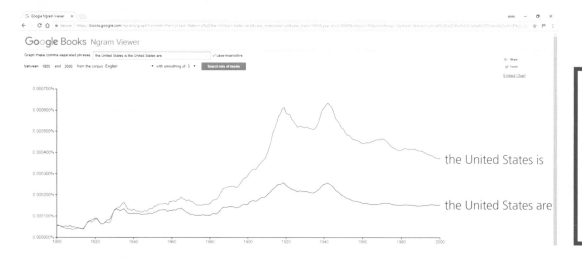

DISCUSS

What is it about his research that qualifies Jean-Baptiste Michel as 'caring'?

How does grouping make probability easier?

You are already very familiar with the 'AND' and 'OR' principle for independent events. Things can get more complicated. For instance, consider this scenario.

A student in Brindisi, Italy, performed a very interesting original (self-composed) song at the school's cultural festival. When asked who her major musical influences were, she responded, 'Pyotr Tchaikovsky and Bob Marley'. Her Mathematics teacher took note, and wondered, 'What are the chances a randomly selected student would listen to such culturally different musicians?' She then surveyed her class.

- S = {all the students in the class}
- Listen to Tchaikovsky: T = {Alessandro, Sofia, Lorenzo, Andrea, Giulia, Aurora, Lucia}
- Listen to Marley: M = {Alessandro, Emma, Elena, Andrea, Riccardo, Lucia}

Can we use the familiar principle of AND → multiply to examine this data?

Does $P(T \cap M) = P(T) \times P(M)$?

Take a close look at the names – why might this not work?

The teacher noticed that a number of students raised their hands twice, once when she asked who listens to Tchaikovsky, and again when she asked who listens to Marley. So, she asked who listens to both:

T ∩ M = {Alessandro, Andrea, Lucia}

Then, she had a student draw a Venn diagram on the board.

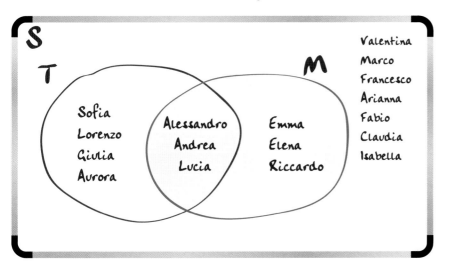

$P(T) = \frac{7}{17}$

$P(M) = \frac{6}{17}$

So,

$P(T) + P(M) = \frac{7}{17} + \frac{6}{17}$

$= \frac{13}{17}$

This does not match! Which answer is correct? Why?

EXAMPLE

Suppose you work for an insurance company and you want to know the probability that a randomly selected policy holder (customer) is considered a 'risky' driver. Your company defines this as someone who has been in an accident **and** who has been driving for less than five years. You have the following information:

- 7000 customers have been in an accident (A)
- 4100 customers have been driving less than five years (F)
- 800 customers are in both groups
- 1700 have more than five years' driving experience and have never been in an accident
- there are 12 000 policyholders in total.

1 The first four numbers do not add up to the last number. Why might this be?
2 What is wrong with this Venn diagram as a representation of the numbers?

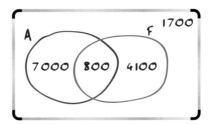

Solution: Problem 1

Consider that all 12 000 policyholders were sitting in a room and a speaker asked those who've been in accidents to raise their hand – the clerk writes down 7000. Next the speaker asks for those drivers with less than five years' experience – the clerk writes down 4100. But 800 people would have raised their hands twice. That means they would be **double counted**!

The students didn't want to copy all this writing, so they simply used numbers to represent how many students belong to each section. This is absolutely okay, in fact preferred by most statisticians as the data are much more readable when dealing with larger groups.

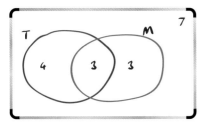

Clearly, the number of desired outcomes is three. We can even name them: Alessandro, Andrea and Lucia. We call this portion of the diagram the **intersection**.

We can see the total number of outcomes if we add up all the names – everyone in the class is listed somewhere in the diagram exactly once.

Then, $\frac{3}{17} = 18\%$

There is an 18% chance that a randomly selected student from the class listens to both musicians.

If we wanted to know the probability of a randomly picked student listening to Tchaikovsky **or** Marley (that is, they raised their hand at least once), we would be interested in the **union** of the two sets.

If we add up all the names in the shaded area, we have 10. So, $P(T \cup M) = \frac{10}{17}$

Can we use our familiar principle of
$P(T \cup M) = P(T) + P(M)$?

Mathematics for the IB MYP 3: *by Concept*

However, we do know the following: 7000 + 4100 + 1700 = 12 800

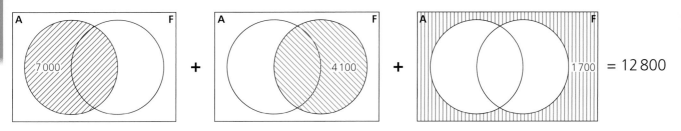

Now we need to subtract one instance of 800 – those who were counted twice (notice the intersection was shaded twice, once in red and once in green).

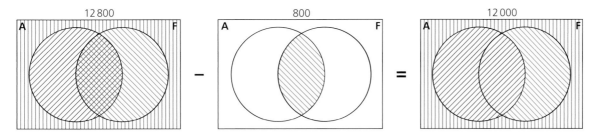

Solution: Problem 2

The Venn diagram has also double counted people in both groups. The best way to go about modelling this is to start in the middle with the overlap and work our way out:

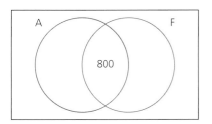

Notice that when we look at the set of policyholders who have been driving less than five years (set F), the 4100 have been subdivided (into 800 and 3300) but are all still within circle F in the Venn diagram.

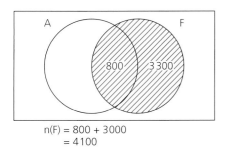

n(F) = 800 + 3000
     = 4100

With the overlap (800) accounted for, we deduct it from each set:

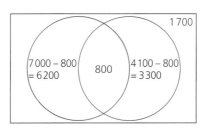

When we look at the set of policyholders who have had accidents (set A), we can see they are divided into those who have had accidents and who have been driving for more than five years (6200), and those who have had accidents and who have been driving for less than five years (800).

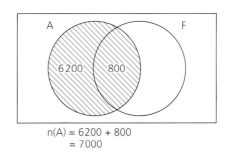

n(A) = 6200 + 800
     = 7000

DISCUSS

Does probability calculated from Venn diagrams fall under theoretical or experimental probability?

Now, finally, we can calculate some probabilities!

3   Find P(x = A) where 'x' represents a randomly selected policyholder. In other words, what is the probability that a randomly selected policyholder has been in an accident?

Solution: Problem 3

Notice that we didn't specify whether or not that policyholder has less than five years' experience. Then we can include both groups within set A. So,

$$P(x = A) = \frac{7000}{12000} = \frac{7}{12}$$

4   P(x = A ∪ F). What is the probability that a randomly selected policyholder has been in an accident **or** has less than five years' driving experience?

Solution: Problem 4

A ∪ F includes anyone in either category. In other words, we are looking for the probability of everyone **except** the policyholders who are in **neither** category.

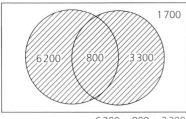

$$P(A \cup F) = \frac{6\,200 + 800 + 3\,300}{12\,000}$$

$$P = \frac{10\,300}{12\,000}$$

$$P = \frac{103}{120}$$

5   If an 'at risk' driver is someone who has had both an accident and less than five years of driving experience, what is the probability of randomly selecting an 'at risk' policyholder?

ACTIVITY: Let's further complicate things

■ ATL

- Creative-thinking skills: Apply existing knowledge to generate new ideas, products or processes

A doctor's office records the symptoms of the first 100 patients of the week.

Symptom	Number of patients
Headache (H)	33
Sore throat (S)	80
Runny nose (R)	68
H ∩ S	30
H ∩ R	6
S ∩ R	50
H ∩ S ∩ R	5

1   Determine n(S).

2   Calculate P(R).

3   Determine the number of patients who had a headache and runny nose but **no** sore throat.

4   Draw a Venn diagram to represent the numbers in the table.

◆ Assessment opportunities

- ◆ In this activity you have practised skills that are assessed using Criterion A: Knowing and understanding.

Solution: Problem 5

The 'at risk' portion of the Venn diagram is the area where the two sets intersect. So,

$$P(A \cap F) = \frac{800}{12\,000} = \frac{1}{15}$$

How can trees serve as metaphors?

Example

Of the 10 000 people in a small town in Thailand, 200 have contracted a newly emergent disease. The whole population is tested for the disease, by a system that has a 1% false positive rate and an 8% false negative rate. This means that of all the tests that give positive results (indicating the disease is present), 1% are wrong (so the disease is not present although the test indicated it was). Of all the negative results, 8% are wrong and these people have the disease.

Calculate the probability that a randomly selected villager has the disease but doesn't realize it!

Solution

This is a lot of information and the question involves calculating multiple probabilities: the first 'event' is the probability of having the disease at all and the second 'event' is the probability of different test results. These types of problems can be even more complex than this example. It is very helpful to learn to organize the information into a tree diagram. Let's set up a diagram for the first 'event' – contracting the disease. We will label all possible outcomes on the diagram, along with their probabilities.

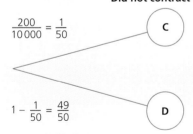

■ Tree diagram for first event, with the possible outcomes on the ends and their associated probabilities along the branches.

Looking at the second 'event', the test result, our possible outcomes are positive and negative. As there is an 8% chance of a false negative, if someone has contracted the disease there is an 8% chance they received a negative result, and a 92% chance (100% − 8%) they tested positive:

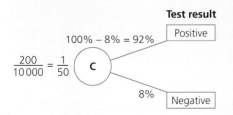

Additionally, if someone does not have the disease, there is still a 1% chance they will test positive (a false positive), and 99% chance they will correctly test negative.

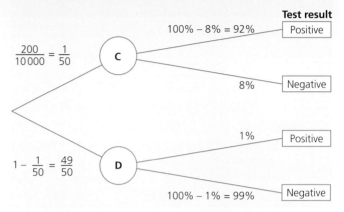

As we travel across various 'pathways' – for instance, 'yes' and 'positive' – we multiply to determine the probabilities of each pair of outcomes.

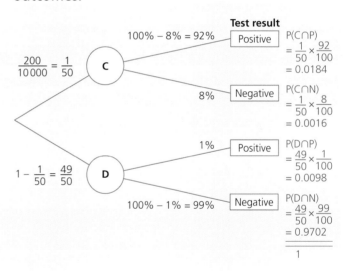

Why is the sum of all four probabilities 1?

Looking at the tree diagram, we can immediately see that, for instance, the probability of having the disease but incorrectly testing negative (and so not realizing the disease is present) is 0.0016.

If we want multiple combinations of outcomes or 'pathways' – for instance the probability of testing negative whether or not the disease is present – we simply add together all the probabilities of 'pathways' that ended in a negative result: 0.0016 + 0.9702 = 0.9718.

Example

Throughout Japan, communities hold Yama, Hoko and Yatai float festivals to pray to the gods for peace and protection from natural disasters.

A local school is asked to randomly select three students from its M3 class to assist in carrying the float. In a school with 14 M3 girls and 18 M3 boys, calculate the probability that:

1 exactly three girls are selected

2 exactly one boy is selected

3 at least one boy is selected.

Solution

Before we begin, let's translate these probabilities into mathematical language:

1 $P(G = 3)$

2 $P(B = 1)$

3 $P(B = 1 \text{ OR } B = 2 \text{ OR } B = 3)$

Let's continue with a tree diagram. If the first pick is a girl, there will one girl fewer to choose from in the second pick, so the probability of picking a girl will decrease. Similar changes to probabilities will result for boys if a boy is picked first, and changes to the third-pick probabilities will likewise depend on the results of the second pick.

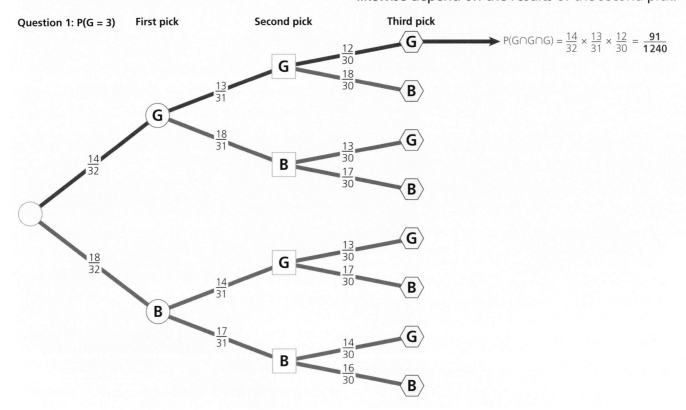

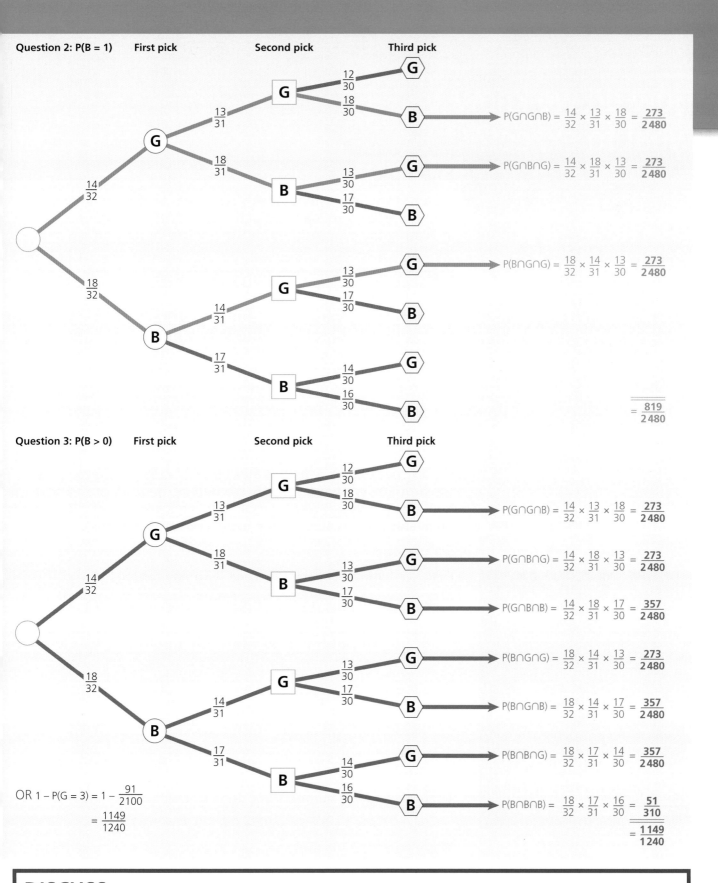

Question 2: P(B = 1)

Question 3: P(B > 0)

$P(G \cap G \cap B) = \frac{14}{32} \times \frac{13}{31} \times \frac{18}{30} = \frac{273}{2\,480}$

$P(G \cap B \cap G) = \frac{14}{32} \times \frac{18}{31} \times \frac{13}{30} = \frac{273}{2\,480}$

$P(B \cap G \cap G) = \frac{18}{32} \times \frac{14}{31} \times \frac{13}{30} = \frac{273}{2\,480}$

$= \frac{819}{2\,480}$

$P(G \cap G \cap B) = \frac{14}{32} \times \frac{13}{31} \times \frac{18}{30} = \frac{273}{2\,480}$

$P(G \cap B \cap G) = \frac{14}{32} \times \frac{18}{31} \times \frac{13}{30} = \frac{273}{2\,480}$

$P(G \cap B \cap B) = \frac{14}{32} \times \frac{18}{31} \times \frac{17}{30} = \frac{357}{2\,480}$

$P(B \cap G \cap G) = \frac{18}{32} \times \frac{14}{31} \times \frac{13}{30} = \frac{273}{2\,480}$

$P(B \cap G \cap B) = \frac{18}{32} \times \frac{14}{31} \times \frac{17}{30} = \frac{357}{2\,480}$

$P(B \cap B \cap G) = \frac{18}{32} \times \frac{17}{31} \times \frac{14}{30} = \frac{357}{2\,480}$

$P(B \cap B \cap B) = \frac{18}{32} \times \frac{17}{31} \times \frac{16}{30} = \frac{51}{310}$

OR $1 - P(G = 3) = 1 - \frac{91}{2100}$

$= \frac{1149}{1240}$

$= \frac{1\,149}{1\,240}$

DISCUSS

How can I recognize when a tree diagram is a good tool to solve a problem?

How could I simulate this problem with a bag of marbles of two different colours?

How do we know what to expect?

EXPECTED VALUE

After we calculate probabilities of quantitative variables, it is often helpful to predict an actual value for those variables. This is a number we can expect to obtain on average. For instance, a help desk may come to expect an average of 14 calls per day. Or a doctor's office may expect to see an average of 27 patients per day. These numbers will vary from day to day, but over many days, weeks or months, they will come close to their **expected value**.

Example

Luka has an inconsistent weekly schedule, so some mornings he is more likely to make his bed than others. Here is a table indicating the probabilities for the number of days he has made his bed in a week.

B = # of days Luka made his bed in a week	P(B)
0	0.02
1	0.07
2	0.1
3	0.21
4	0.23
5	0.16
6	0.12
7	0.09

Notice that the probabilities add up to 1, so this table includes the probability of every possible outcome. If we want to know how many days per week we can **expect** Luka to make his bed – on average – we calculate the expected value of B, E(B).

Solution

First, we'll multiply each outcome with its probability:

B = # of days Luka made his bed in a week	P(B)	B*P(B)
0	0.02	0
1	0.07	0.07
2	0.1	0.2
3	0.21	0.63
4	0.23	0.92
5	0.16	0.8
6	0.12	0.72
7	0.09	0.63

Then we add the numbers in our new column:

$$0 + 0.07 + 0.2 + 0.63 + 0.92 + 0.8 + 0.72 + 0.63 = 3.97$$

This means that, on average, we can expect Luka to make his bed nearly four times per week. Some weeks may be more, some may be less, but in the long run it will work out to almost four.

This was a 'clean' example, as the probabilities were already given to us. In most cases, the probabilities would need to be calculated before determining their expected value.

Example

Determine the expected number of hours of sunshine on a random day in Romania based on the information from weather-and-climate.com.

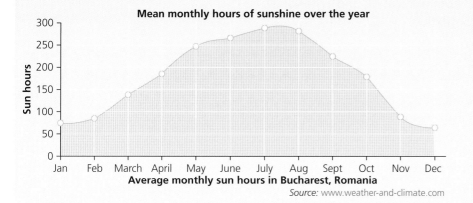

Mean monthly hours of sunshine over the year

Sun hours (y-axis): 0, 50, 100, 150, 200, 250, 300

Months (x-axis): Jan, Feb, March, April, May, June, July, Aug, Sept, Oct, Nov, Dec

Average monthly sun hours in Bucharest, Romania

Source: www.weather-and-climate.com

Solution

Let's begin by calculating the probability of sunshine hours in January. January has 31 days, and thus $31 \times 24 = 744$ hours.

There were 75 hours of sunshine out of 744 total hours in January, so the probability of a sunny hour is $\frac{75}{744}$.

Copy and complete the table for the rest of the months of the year.

Month	H (# of hours of sunshine)	P(H)
Jan	75	
Feb		
March		
April …		

MEET A MATHEMATICIAN: YOU!

Learner Profile: Risk-taker

'Math is being able to engage in joyful intellectual play – and being willing to flail (even fail!).'

James Tanton

Do you ask questions in class **every** time something is unclear? If you have ideas that others might think are really obscure and odd, do you voice them? If not, what is your greatest fear in performing these actions? Write it down; it will come in handy when you take action later.

The most brilliant mathematicians fail **most** of the time. They fail because they are trying to solve really difficult problems that nobody before them has ever solved. In failing, they are still making important discoveries. In order to solve a problem that has been stumping everyone else, mathematicians need to try new approaches – even if they are far-fetched and seem likely to fail. The only solution that will work is a brand new one, so go for it!

This applies to mathematicians of all ages, even you! Perhaps someone before you may have already discovered a formula or noticed a relationship, but if you are discovering it yourself, you are reinventing mathematics – creating it afresh. So have courage, take risks – encourage your curious brain and ask someone to fill in the blanks as they come up. They come up for **everyone** at some point or another!

SUMMATIVE ASSESSMENT: Chance carnival

Use these problems to apply and extend your learning in this chapter. The problems are designed so that you can evaluate your learning at different levels of achievement in Criterion C: Communicating and Criterion D: Applying mathematics in real-life contexts.

Your group's task is to design a unique (new, not one that already exists) **carnival**, **amusement park** or **casino**-style game. You must be able to **explain** the probability of your game, so don't make it too complicated!

The carnival (one submission per group)

1 **The game** – create all the essential parts of the game (game boards, playing pieces, cards, balls, etc.) to be played at the carnival.
2 **Game description** – this should set out what you need to play the game.
    - List all materials needed to play (dice, spinner, darts, ball, for example).
    - Draw a picture of your game board.
3 **Introduction** – provide an overview of your game.
    - What type of game is it?
    - Where would you play this type of game?
    - How much does it cost to play?
    - What are the prizes if you win?
4 **Instructions** – create a set of step-by-step instructions which explain your game clearly enough that anyone can pick them up, read them and begin playing.

The write-up (one submission per student)

1 **Probability analysis** – describe the aspects of probability used in your game.
    - Is the game fair? Show the mathematical calculations for the expected value of winning the game.
    - If the game is not fair, how could you change it to make it fair?
2 **Reflection** – write a one-page reflection.
    - What were your overall feelings about this project?
    - Did this project help you understand the topic of probability any better?
    - How did your group work together?
    - Were there any group members who didn't pull their weight? Any group members who tried to boss the group around?
    - How successful was your game?
    - What have you learned about 'fair games'?
    - What is your opinion about the gambling industry in general?
    - Do you think it's fair to have establishments that are designed for people to lose money?
    - Would you advise your grandmother to play the national lottery? Why/why not?
    - In which religions and cultures is gambling unacceptable? Why might this be?

Reflection

Questions we asked	Answers we found	Any further questions now?			
Factual: How do we know what to expect?					
Conceptual: Are all probabilities created equal? Why does mathematics sometimes overcomplicate simple problems? How does grouping make probability easier? How can trees serve as metaphors?					
Debatable: Is probability just for fun? Does all probability have to be theoretical?					
Approaches to Learning you used in this chapter:	Description – what new skills did you learn?	How well did you master the skills?			
		Novice	Learner	Practitioner	Expert
Communication skills					
Information literacy skills					
Creative-thinking skills					
Critical-thinking skills					
Learner Profile attribute(s)	Reflect on the importance of being caring and a risk-taker in this chapter.				
Risk-taker					
Caring					

5 Where's the proof?

Finding **relationships** in closed **systems** can help us **simplify** and solve problems, using **technology** or otherwise.

CONSIDER THESE QUESTIONS:

Factual: What is trigonometry? What is Pythagoras' theorem? What are trig ratios? How can we find any 'part' of any right-angled triangle given two values?

Conceptual: How do geometric relationships appear in real life? Is simplification always useful? What is meant by a closed system?

Debatable: If trigonometry can help us find unknowns, how much can we trust this in different systems and in different situations? Do trigonometric abstractions always give us the right answer in all situations? Does innovation create new mathematical relationships or help us to discover old ones in new ways?

Now **share and compare** your thoughts and ideas with your partner, or with the whole class.

ΠΥΘΑΓΟΡΑΣ Ο ΣΑΜΙΟΣ
580 – 496 π.χ

IN THIS CHAPTER, WE WILL ...

- **Find out** how shapes' lengths and angles depend on each other.
- **Explore** why a theorem is a fundamental idea in mathematics.
- **Take action** by looking for answers to 'out of reach' problems.

THINK-PAIR-SHARE

What do these triangles have in common? What are they trying to communicate?
How are they different from one another? What does it make you think?

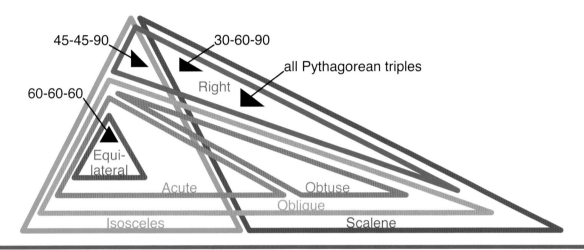

■ These Approaches to Learning (ATL)
skills will be useful …

- Communication skills
- Creative-thinking skills
- Transfer skills
- Critical-thinking skills

● We will reflect on this Learner Profile
attribute …

- **Thinker** – We use critical- and creative-thinking
  skills to analyse and take responsible action
  on complex problems. We exercise initiative in
  making reasoned, ethical decisions.

◆ Assessment opportunities in
this chapter:

- **Criterion A:** Knowing and understanding
- **Criterion B:** Investigating patterns
- **Criterion C:** Communicating
- **Criterion D:** Applying mathematics in real-life
  contexts

PRIOR KNOWLEDGE

Reflect on what you already know about:
- the names of certain shapes (polygons),
  especially different types of triangles
- square numbers and square roots
- how to measure and draw line segments of
  certain lengths
- the sum of all angles in a triangle

KEY WORDS

adjacent	bird's eye view
opposite	decimal places (d.p.)
right-angled	

What is Pythagoras' theorem?

WHAT IS A THEOREM?

ⓘ **Theorem**:
A statement that can be demonstrated to be true by accepted mathematical operations and arguments.

In general, a theorem is evidence of some general principle that makes it part of a larger theory. These theorems build up the body of knowledge within geometry and mathematics. The process of showing a theorem to be correct is called a **proof**.

Proving theorems is a valuable skill and can help you to improve your logic, reasoning and ability to identify things that are indisputably, and in every case, true. When you are learning a theorem, you are learning an absolute truth, which will help you to discover other truths.

How can we use theorems? How do mathematicians find theorems?

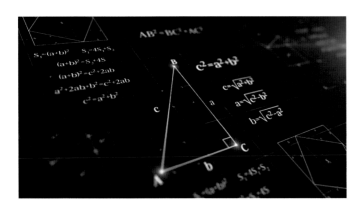

MEET A MATHEMATICIAN: PYTHAGORAS, (C.580BCE–C.500BCE)

Learner Profile: Thinker

Pythagoras was an Ancient Ionian Greek philosopher and mathematician who is estimated to have lived between 570 and approximately 495BCE. He is said to have made contributions to the fields of religion, music and astronomy, and is credited as the founder of the movement called Pythagoreanism.

The exact details of Pythagoras' life and works have been lost in the mists of time, and much of what we know about him was written down centuries after his lifetime. He is most commonly remembered for **Pythagoras' theorem** which is perhaps one of the oldest and most elegant relationships in geometry.

Many of Pythagoras' followers worked on a variety of mathematical relationships and were very secretive about their work and discoveries. In the absence of reliable information, the theorem has been credited to Pythagoras but it is also possible that it was proven by one of his followers. There is some evidence that shows many other, much older, civilizations were aware of the relationship before Pythagoras' time, although this is the oldest, or best known, **formal proof**.

EXTENSION

Research details of the Pythagorean movement. Who were they? What did they believe? What does this tell you about the origins of mathematics and philosophy?

Any theorem, no matter how difficult to prove in the first place, is viewed as 'trivial' by mathematicians **once** it has been proven.

Therefore, there are exactly two types of mathematical objects: trivial ones, and those which have not yet been proven.

A mathematician is a machine for converting coffee into theorems.

Weak coffee was suitable only for lemmas (a short theorem used in proving a larger theorem).

- Nobel Prize-winning physicist Richard Feynman
- P. Erdős
- Paul Turán

What is Pythagoras' theorem?

Remember that:

A right-angled triangle contains a right angle.

The longest side opposite the right angle is called the hypotenuse.

One way to find the length of the hypotenuse is to measure it. This is a possible solution but it:
- might not be possible to measure it in **real life**
- won't tell you anything about other right-angled triangles in general or in **abstract**.

Pythagoras' theorem states that the square of the hypotenuse is equal to the square of the other two sides. This can be stated using algebra notation:

$a^2 + b^2 = c^2$ where a and b are the shorter sides and c is the hypotenuse

Now we are going to prove this theorem for all cases (for all right-angled triangles). The relationship between the sides is not obvious if we look at the lengths, or measure in one dimension (1D).

It is clear from the diagram that simply adding 6 to 8 does not equal 10 mathematically, represented as:

$$6 + 8 \neq 10$$

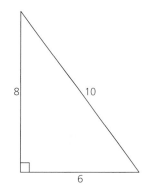

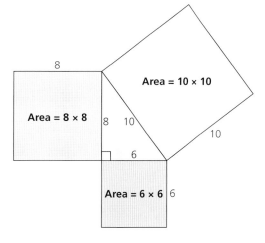

By squaring each side, we see that the sum of the shorter sides squared is 100:

$$6^2 + 8^2 = 36 + 64$$
$$= 100$$

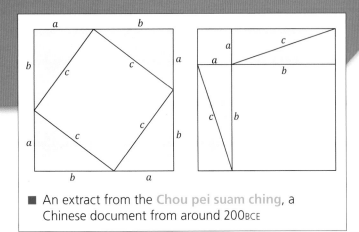

■ An extract from the Chou pei suam ching, a
Chinese document from around 200BCE

If we compare this to $10^2 = 100$, we can say the
theorem is proven. Now we must remove the numbers
and replace these with algebraic notation (letters) to
see if it holds true for every case.

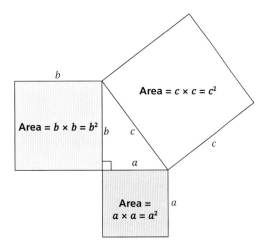

By showing that the areas of both smaller squares
added together equals the area of the largest square,
we have proven the relationship for **any** values in a
right-angled triangle.

Visual proofs such as these are powerful and
appropriate ways to demonstrate a theorem in a direct
and accessible way. Other visual proofs for Pythagoras'
theorem from history include:

- **www.youtube.com/watch?v=CAkMUdeB06o**
- **en.wikipedia.org/wiki/Pythagorean_theorem**

EXTENSION

One of the proofs of the theorem is to show
that the sum of the area of the squares on
each side equals the area of the square on
the hypotenuse, like this:

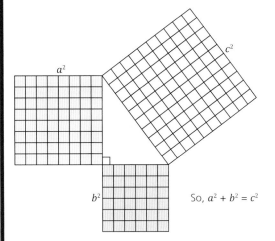

So, $a^2 + b^2 = c^2$

Would it work for other shapes placed on the
sides of the triangle? Investigate this idea to
see if it holds true for other shapes.

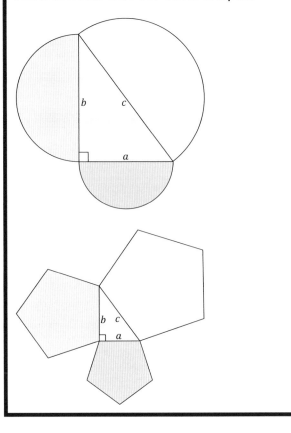

HOW DO WE USE PYTHAGORAS' THEOREM TO FIND LENGTHS?

ACTIVITY: Hunt the hypotenuse

ATL

■ Communication skills: Use and interpret a range of discipline-specific terms and symbols

Find and label the hypotenuse in each of the shapes shown here. How many hypotenuses can you find?

1 to 3 hypotenuses	6 to 7 hypotenuses
4 to 5 hypotenuses	8 to 10 hypotenuses

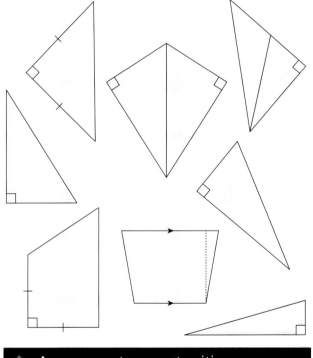

◆ Assessment opportunities

◆ In this activity you have practised skills that are assessed using Criterion A: Knowing and understanding.

Examples

Now that we have practised finding hypotenuses and seen the theorem, it is time to apply it.

1

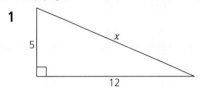

Solution

$$a^2 + b^2 = c^2$$
$$5^2 + 12^2 = x^2$$
$$25 + 144 = x^2$$
$$169 = x^2$$
$$x^2 = 169$$
$$x = \sqrt{169}$$
$$x = 13$$

> **Hint**
>
> Don't forget units, if given in the question.

2

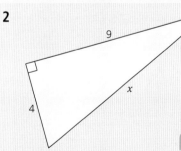

Solution

$$a^2 + b^2 = c^2$$
$$9^2 + 4^2 = x^2$$
$$81 + 16 = x^2$$
$$97 = x^2$$
$$x^2 = 97$$
$$x = \sqrt{97}$$
$$x = 9.848857802$$
$$x \approx 9.8$$

> **Hint**
>
> Were you asked for a certain number of decimal places (d.p.) or significant figures (s.f.)? Do you have to choose your own accuracy? Always check whether this is specified in the question, so you know how to present your answer at the end.

PRACTICE EXERCISE

1 Find the value of x.

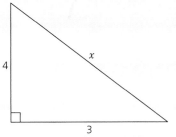

2 Find the value of f.

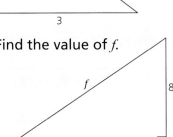

3 Find the length of the hypotenuse in each case. Give your answer correct to 1 d.p.

a

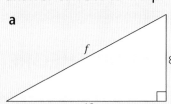

b

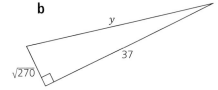

4 Find the length of the hypotenuse in each case, leaving your answers in root (or surd) form.

a

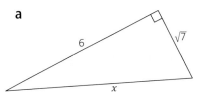

b

3 Given a height of 150 m and a distance along the ground of 100 m for a zipline, as pictured, how long is the portion of zipline, z?

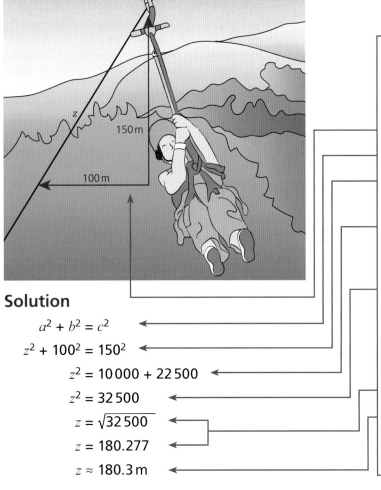

Solution

$$a^2 + b^2 = c^2$$
$$z^2 + 100^2 = 150^2$$
$$z^2 = 10\,000 + 22\,500$$
$$z^2 = 32\,500$$
$$z = \sqrt{32\,500}$$
$$z = 180.277$$
$$z \approx 180.3 \text{ m}$$

Procedure pro: a how-to guide

This is a good method to help you to carry out and communicate your working!

1 Label the sides.

2 Write the formula down.

3 Substitute the values correctly. (Make sure to check!)

4 Complete the calculations. (Use a calculator if appropriate, but don't get lazy.)

5 Show interim values in your calculation, so a reader can see what numbers you got before finding the square root (and if you get later steps wrong, you will still get some credit).

6 Find the square root of the sum.

7 Give the answer to the appropriate number of s.f. (Look at the question for guidance or use your common sense.)

What about the other sides?

USING PYTHAGORAS' THEOREM TO FIND THE SHORTER SIDES

We can use Pythagoras' theorem to help find short sides too, provided we know the length of the hypotenuse. Algebraic rearranging, as you learned in *Mathematics for the IB MYP 1* and *2*, allows you to move operations and reorder equations whenever necessary.

Watch what happens when we do not know one of the short sides, but we do have the hypotenuse.

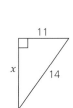

$$a^2 + b^2 = c^2$$

$$x^2 + 11^2 = 14^2$$

$$x^2 + 121 = 196$$

$$x^2 = 196 - 121 \quad \text{Notice how we}$$
$$\qquad\qquad\qquad\qquad \text{subtract here.}$$

$$x^2 = 75$$

$$x = \sqrt{75}$$

$$x \approx 8.7$$

Now with a different example:

$$a^2 + b^2 = c^2$$

$$x^2 + 7^2 = 22.5^2$$

$$x^2 = 22.5^2 - 7^2$$

$$x^2 = 506.25 - 49$$

$$x^2 = 457.25$$

$$x = \sqrt{457.25}$$

$$x = 21.38340478\,\text{m}$$

$$x \approx 21.4\,\text{m}$$

PRACTICE EXERCISE

Find the missing side in each case.

1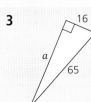

2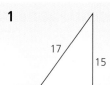

3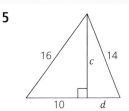

4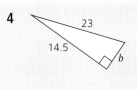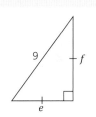

5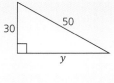

6

THINK-PAIR-SHARE

Using Pythagoras' theorem, are these triangles right-angled or not? How can you tell? How can you prove this? You may measure the lengths of the sides of the triangles, if you wish. They are drawn to scale.

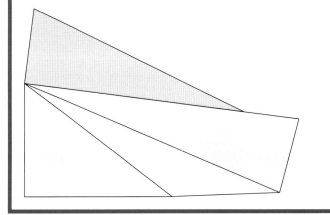

Example: A word problem

In September 2015 Instagram announced that they would be enabling upload of landscape images, in addition to the traditional square format. The maximum size for this type of upload would be 450 × 600 pixels. What is the length of the diagonal of this new format?

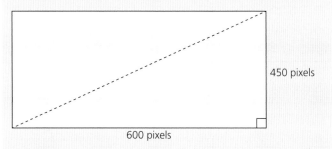

450 pixels

600 pixels

Solution

$$450^2 + 600^2 = \text{diagonal } (d)^2$$
$$d^2 = 202\,500 + 360\,000$$
$$d^2 = 562\,500$$
$$d = \sqrt{562\,500}$$
$$d = 750 \text{ pixels}$$

HOW CAN WE SOLVE PROBLEMS WITH PYTHAGORAS' THEOREM?

Triangles, particularly right-angled triangles, are present all around us in the three-dimensional space of our everyday lives. Many problems can be solved using Pythagoras' theorem, although some problems may not be immediately obvious as triangular in nature.

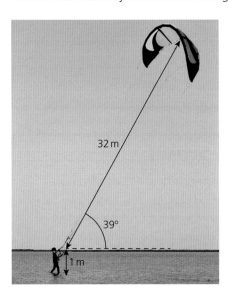

32 m

39°

1 m

For example, the length of the kite string and the horizontal distance from the holder (along the ground) can be used to find the height at which the kite is flying.

How do I find the hidden triangles?

It is very helpful to imagine the triangle 'hidden' in the problem. First look for the lengths or measurements you have been given and mark them on the image. Remember to mark the right angle, both to show your communication and to check that you have correctly identified the hypotenuse.

PRACTICE EXERCISE

Find the missing length in each question. State your answers correct to 2 d.p.

1 Find the length of the side of the cone.

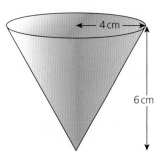

4 cm

6 cm

2 Find the length of the staircase.

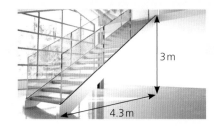

3 m

4.3 m

EXTENSION

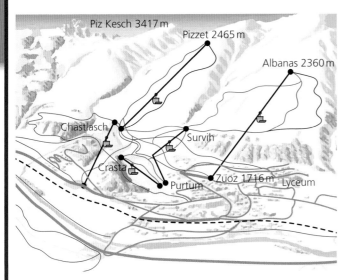

Piz Kesch 3417 m
Pizzet 2465 m
Albanas 2360 m
Chastlasch
Survih
Crasta
Purtum
Zuoz 1716 m
Lyceum

Let's consider the length of a ski run on this Swiss mountain.

You want to find the distance you will have skied between Albanas and Zuoz. You know that the distance between Zuoz and Albanas on a map is 500 m but this is the distance as seen from above (a bird's eye view). How do you find a triangle in that?

Given that you know both of the towns' heights above sea level, you can calculate the distance between their heights. But you didn't ski down through the Earth! Your route was far longer and gentler than that!

You need to visualize and 'extract' a right-angled triangle to help you to find the length of the piste.

Again identify the lengths. First, the distance between the towns is 500 m. Next the difference in height (or altitude) can be found by subtracting. And we are looking for the length of the slope or piste.

Imagine you could slice into the mountain and look at the cross-section shown in the diagram.

Therefore,

$500^2 + 644^2 = 664\,736$

So, the length of the ski run must be:

$\sqrt{664\,736}$

$\approx 815\,\text{m}$

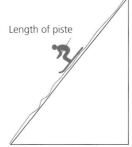

Length of piste

Difference in altitude
2360 − 1716
= 644 m

Distance on a map 500 m

---

3 Find the height of the ramp, where the total length along the ramp is 12 m and the base of the ramp is 9 m.

4 Find the length of the base of the triangle shown.

2 m
7 m

5 Find the length of the step side of the ladder.

2 m
1 m

6 Find the length of the escalator.

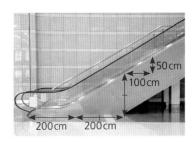

50 cm
100 cm
200 cm   200 cm

ACTIVITY: Find the fake

A Pythagorean triple is defined as measurements of three sides which are whole-number solutions for the sides of a right-angled triangle. The simplest one is 3, 4, 5 because

$$3^2 + 4^2 = 5^2$$

1 Prove that the triple (5, 12, 13) is a Pythagorean triple.
2 Prove that the triple (20, 21, 27) is not a Pythagorean triple.
3 For the following list of triples, find the fakes.
   a (8, 15, 17)   e (280, 450, 533)
   b (7, 2, 25)   f (110, 60, 651)
   c (12, 35, 37)   g (16, 63, 65)
   d (19, 40, 41)   h (3300, 5600, 6500)

4 Choose a triple that you know is Pythagorean. Predict whether enlarging (or scaling up) each side by the same amount will mean it is no longer Pythagorean.
5 Now, try it out:
   a Multiply each side by 2 and verify Pythagoras' theorem with the new numbers.
   b Multiply each side by 10 and verify Pythagoras' theorem with new numbers.
   c Halve each side and verify if it is still a triple.
   d What observation or general rule can you make from this? Justify your answer.

Can you find a Pythagorean triple that consists of three even numbers? Three odd numbers?

WHAT MAKES YOU SAY THAT?

Look at the image and discuss the following questions with a partner.

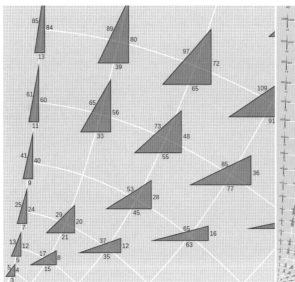

1 What's going on?
2 What do you see that makes you say that?

This image shows Pythagorean triples plotted on a **curvilinear grid**, which is composed of curves instead of straight lines. The odd leg of the triangle is plotted on the horizontal curves and the even leg of the triangle is plotted on the vertical curves.

Describe how the triples relate to one another. How are they laid out? How do the sides change? What patterns can you see?

ACTIVITY: The Pythagoras tree

ATL

- Creative-thinking skills: Create original works and ideas; Use existing works and ideas in new ways

The Pythagoras tree was first constructed in 1942 by Albert E. Bosman (1891–1961), a Dutch mathematics teacher.

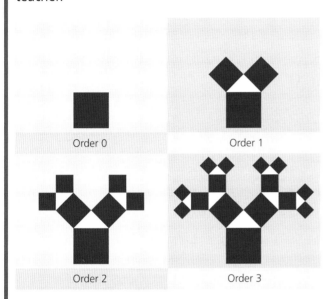

Order 0   Order 1

Order 2   Order 3

Part A

To start the tree, we begin with a box of equal sides. On top of one side, we place two squares that are angled until they make an isosceles right-angled triangle (in white).

Copy and complete the following table.

Order number	Number of squares (purple)	Number of right-angled triangles (white)
0		
1		
2		
3		

1. Can you identify any pattern(s)?
2. Express this pattern as a general rule.
3. Verify and justify your answer.

Part B

Use a large piece of square paper. Complete the tree to order number 6, making the first square 10 × 10 boxes. Shade in the triangles in a different colour.

1. What do you notice?
2. At which order do you think the shapes will be too small to see?
3. What is the size of the entire tree, relative to your grid?

EXTENSION

The smaller squares are always reduced or scaled down by the same factor. What is that factor?

◆ Assessment opportunities

- In this activity you have practised skills that are assessed using Criterion A: Knowing and understanding, Criterion B: Investigating patterns and Criterion C: Communicating.

HOW CAN WE FIND UNKNOWN ANGLES?

Trigonometry, which comes from the Greek meaning 'triangle measurement', uses the property of similarity to find unknown sides or angles. The fact that side lengths of similar triangles are always in the same ratio has allowed mathematicians to name these ratios and devise uses for them.

These three ratios, **sine**, **cosine** and **tangent**, each use two of the three sides of a right-angled triangle, relative to an angle.

First, we need to be able to identify and label the sides of a triangle correctly. We know from learning about Pythagoras' theorem that the hypotenuse is the longest line, **always** opposite the right angle. The other two sides of a right-angled triangle are called **opposite** and **adjacent** (next to) according to their position relative to the angle in question.

REFLECTION

Why are these ratios always the same, no matter what the size of the triangles?

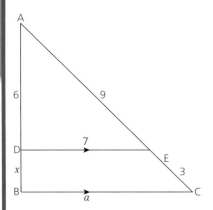

Measuring angles? Can you use a protractor? What about measuring large angles or angles in real life?

Find out what a clinometer is and how it works. You could make some homemade ones for school as a Service activity.

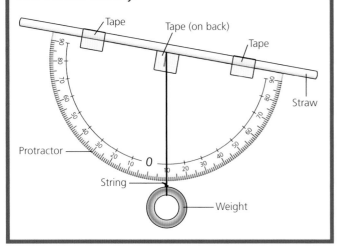

Once an angle, θ, is specified, we can identify the adjacent side (the side next to the angle) and mark it with an **A**, like this:

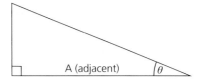

Next we identify the side opposite the angle with a large **O**, like this:

For good communication your letter O must look like a letter O and not a zero, 0.

If we start at a different angle, then the names of the sides must change relative to the new angle.

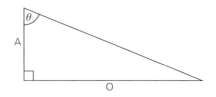

PRACTICE EXERCISE

1 **Copy and complete the missing labels for this triangle.**

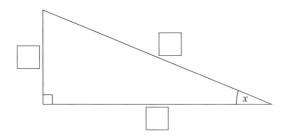

2 **Given these labels for the sides, find the angle they relate to.**

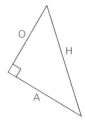

3 **If the ratio of two sides always gives us a constant number, there can only be three possible pairings made from adjacent, opposite and hypotenuse. List all the possible combinations of pairs.**

How do we use the trigonometric ratios to find lengths?

Mathematics hack – SOHCAHTOA

This is a tool that can be used to help remember the order of the ratios and how to find them. By laying them out in this way, you can easily select the correct ratio for your needs and you are reminded how to find the **quotient** (which number is the numerator and which is the denominator).

In the last exercise you stated all the possible combinations of triangle sides. Now it's time to look at those pairings as individual ratios or quotients.

The pairing of opposite with hypotenuse has been given the name **sine** (shortened to sin) where:

$\sin \theta = $ opposite $\div$ hypotenuse

Similarly, the ratio between adjacent and hypotenuse has been given the convention **cosine** (shortened to cos), where:

$\cos \theta = $ adjacent $\div$ hypotenuse

The final possible combination is opposite and adjacent. This ratio is called **tangent** (abbreviated to tan) and is stated as:

$\tan \theta = $ opposite $\div$ adjacent

$\underset{\text{S = O} \div \text{H}}{S\,^O_H} \quad \underset{\text{C = A} \div \text{H}}{C\,^A_H} \quad \underset{\text{T = O} \div \text{A}}{T\,^O_A}$

For a right-angled triangle, the sine, cosine and tangent of the angle θ are defined as:

$\sin \theta = \frac{\text{opposite}}{\text{hypotenuse}} \qquad \cos \theta = \frac{\text{adjacent}}{\text{hypotenuse}} \qquad \tan \theta = \frac{\text{opposite}}{\text{adjacent}}$

Sin θ will always have the same value for any particular angle, regardless of the size of the triangle.

Example

Find the value of O.

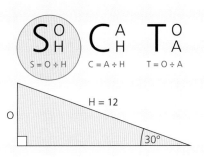

Solution

1. Label sides.
2. Write SOHCAHTOA.
3. Decide on and circle the best ratio to use.
4. Write the formula.

   $\sin 30° = \frac{\text{opposite}}{\text{hypotenuse}}$

5. Substitute correctly.

   $\sin 30° = \frac{O}{12}$

6. Rearrange to solve.

   $O = 12 \times \sin 30°$

7. Calculate the answer and give units if appropriate.

   $O = 6$

Examples

1. Use your calculator to find the following ratios.

   a $\sin 45°$

   b $\cos 87°$

   c $\tan 22°$

Solution

   a $\sin 45° = 0.707$

   b $\cos 87° = 0.052$

   c $\tan 22° = 0.404$

2 Find the length marked x.

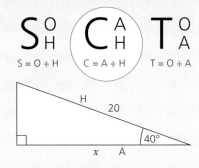

$S = O \div H$ $C = A \div H$ $T = O \div A$

Solution

$\cos 40° = \dfrac{\text{adjacent}}{\text{hypotenuse}}$

$\cos 40° = \dfrac{x}{20}$

$x = 20 \times \cos 40°$

$x = 15.3$

3 The triangle above the front door in this picture is an isosceles triangle with base angles of 25° and a width of 4 m. How tall is the triangle?

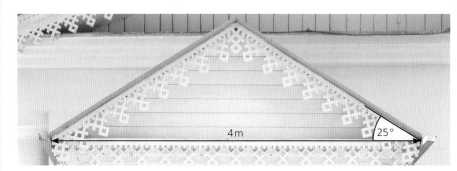

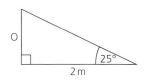

Solution

$\tan 25° = \dfrac{O}{2}$

$O = 2 \tan 25°$

$O = 0.933\,\text{m}$

PRACTICE EXERCISE

1 Using your calculator, find sin 13°, cos 31°, tan 52°, tan 91°, tan 44°, cos 44°, sin 58°.

2 Label the sides O, A or H as appropriate on these triangles.

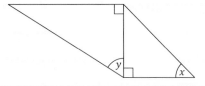

3 Find the height of the stairs, given the information in the image.

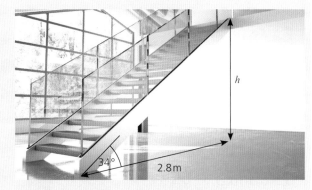

4 Sketch a triangle with an angle of 45° which has an opposite side of 4 cm.

5 Given the dimensions in question 4, find:

  a the adjacent side

  b the hypotenuse

  c the other two angles.

6 What assumption(s) did you make in questions 4 and 5? How did it affect your answer?

If you are working with sin or cos and you find the number below the line (the denominator) is smaller than the number above the line (the numerator), **stop**! This combination will lead to an error. The calculator will not be able to find it. Check your working! Maybe you mislabelled the sides.

REFLECTION

Why is this the case? Why can't the denominator be smaller than the numerator? Is this also true for tan?

ACTIVITY: Graphing with dice

■ Communication skills: Make inferences and draw conclusions

1 Using your calculator, find a value for the sine of each angle in the table.

θ (angle)	$\sin \theta$
0°	
30°	
60°	
90°	

2 What pattern(s) can you see? Can you use the pattern to predict another sine, perhaps sin 20° or sin 50°?

3 Copy and complete the following table with additional values for sine.

θ (angle)	$\sin \theta$
0°	
10°	
20°	
30°	
40°	
50°	
60°	
70°	
80°	
90°	

4 How closely did your pattern match the results in the table? Why might this be?

5 Do you think the other ratios work in the same way?
Predict how you think the pattern will work for cosine (cos) and tangent (tan) of the angles between 0° and 90°.

6 Copy and complete the following table, with additional values for cosine (cos) and tangent (tan).

θ (angle)	$\sin \theta$	$\cos \theta$	$\tan \theta$
0°			
10°			
20°			
30°			
40°			
50°			
60°			
70°			
80°			
90°			

7 What patterns can you see in the table?

8 How closely did your pattern (or prediction) from question 5 match the results of the table? Why might this be?

9 How does that connect to what you already know about sin, cos and tan?

10 For each of the ratios, plot the values from the table on a graph, placing the angles from 0° to 90° on the x-axis (horizontal).

11 Use two dice to generate some random values to calculate and plot on the graph you completed for question 10. For example, if you roll a 3 and a 2, find sin 32°, cos 32° and tan 32° using your calculator. Then mark the values on the graph.

12 Comment on what you have learned about sin, cos and tan from this activity.

EXTENSION

Add a column to your table to show the quotient (division) of $\dfrac{\sin \theta}{\cos \theta}$.
What do you notice? Why is this?

◆ Assessment opportunities

◆ In this activity you have practised skills that are assessed using Criterion B: Investigating patterns.

How do we use the trigonometric ratios to find angles?

So far, we have used ratios to find a missing side in a right-angled triangle. What if it is an angle that is unknown?

In algebra you learned rules for rearranging formulae to find an unknown. In the previous procedures, some inversing was required (squaring and square rooting) to find the answer. Trigonometric ratios also each have an inverse function.

- $\sin \to \sin^{-1}$
- $\cos \to \cos^{-1}$
- $\tan \to \tan^{-1}$

If we know the value of a sine (ratio of two sides) in a given triangle is $\sin x = 0.5$, we need to isolate or solve for x to find the angle. To remove the sine (or to move it to the right-hand side of the expression), we must use the inverse: $\sin^{-1}$.

To work this out we use the $\sin^{-1}$ key on the calculator:

$$\sin^{-1} 0.5 = 30°$$

$\sin^{-1}$ is the inverse of sin. It is sometimes called **arcsin**.

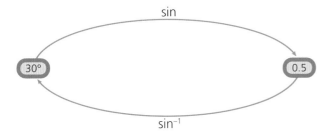

Examples

Find the missing angle x in each case.

1

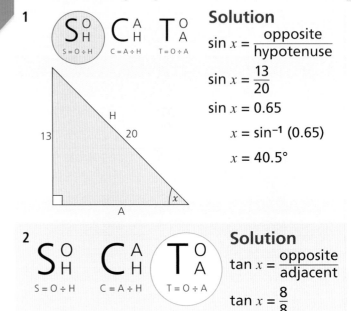

Solution

$$\sin x = \frac{\text{opposite}}{\text{hypotenuse}}$$

$$\sin x = \frac{13}{20}$$

$$\sin x = 0.65$$

$$x = \sin^{-1}(0.65)$$

$$x = 40.5°$$

2

Solution

$$\tan x = \frac{\text{opposite}}{\text{adjacent}}$$

$$\tan x = \frac{8}{8}$$

$$x = \tan^{-1}(1)$$

$$x = 45°$$

Using your calculator or graphing calculator:

You already know that scientific calculators have sin, cos and tan buttons. Most apps or calculators also have inverse buttons. So, press the 'inverse' or 'shift' button followed by the 'sin' button for 'sin⁻¹', and so on. Ensure that your calculator is in degrees mode and not radian, which is very important. You will learn more about radians in later years.

3 If the step side of the ladder is 4 m long and the distance between the sides of the ladder is 1.5 m, what angle does the ladder make with the ground?

$$S^O_H \quad \widehat{C^A_H} \quad T^O_A$$

$$S = O \div H \qquad C = A \div H \qquad T = O \div A$$

Solution

$$\cos x = \frac{\text{adjacent}}{\text{hypotenuse}}$$

$$\cos x = \frac{0.75}{4}$$

$$\cos x = 0.1875$$

$$x = \cos^{-1}(0.1875)$$

$$x = 79.2°$$

PRACTICE EXERCISE

1 Find the angle the ramp makes with the ground.

2 If the height of the box is 3 m and the ramp length is 5 m, find the other side of the triangle. Find all three angles of the triangle.

Discuss one cause of inaccuracy or an assumption you made which could affect your result.

3 Martha and Kate are buying a 6-foot Christmas tree. The problem is that their car is only 4 feet long in the back. At what angle should they place the tree to fit it in the car?

Estimate whether, with an internal height of 1 m, the car is tall enough to get the tree home.

4 Matt thinks that each diagonal of the screen splits the screen into the same angles every time. Investigate to see if he is correct or incorrect. You can measure or research to find the screen dimensions.

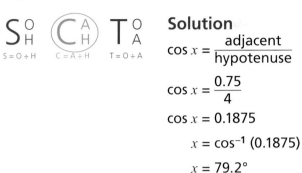

How do geometric relationships appear in real life?

FORMATIVE ASSESSMENT EXTENSION

Using Pythagoras in three dimensions

You have seen this lovely idea for a hammock (swing) on Pinterest and your family has asked you to calculate all the dimensions to buy the correct materials to make it.

Your parents want it large enough for adults and children to use. They would like it to look like the image shown, but the exact dimensions and design of the hammock are up to you.

In your garden, there are two trees that are strong enough and they grow almost horizontally in parts, similar to the palm tree in the picture. One tree is 1.5 m tall and the other is at a height of 2.5 m.

Your task is to create a plan for the construction of this hammock. You must include the following:

- a detailed drawing of the hammock
- a list of materials required
- all dimensions of the individual parts and any other measurements you think you will need
- the total length of rope required.

1 Using the information you have learned in this chapter, identify all the relevant elements for the design and calculate the length of rope needed for your chosen size and height of hammock (strands i to iii).

2 Once you have completed the calculations and design drawing, you should comment on the accuracy of your calculations and whether your final product makes sense in terms of the context of the problem (strands iv to v).

◆ Assessment opportunities

◆ In this activity you have practised skills that are assessed using Criterion D: Applying mathematics in real-life contexts.

ACTIVITY: Would you look at that house?

■ ATL

- ■ Transfer skills: Inquire in different contexts to gain a different perspective

You are reading an article online about unlikely houses and see this image. The house is clearly tilted at an angle and your task is to find a way to calculate the angle of tilt.

Treat this task as if you were actually at the location in real life and are able to make all the necessary measurements to find that unknown angle.

Part A: Leaning house

Given this scenario, take the following steps to find the angle of tilt for the house.

1 **Sketch the house on a piece of paper, leaving enough room to make notes around it.**
2 **Identify the right-angled triangle you will use to make your calculations.**

3 **Describe the measurements you would take and how you would make them.**
4 **Estimate (make up) sensible dimensions for these measurements and explain how you decided on them (using mathematical reasoning).**
5 **Using the values you have estimated, show how you could calculate the angle of tilt.**

Part B: Quick change

You search online and find another image of the same house, similar to the previous one in some ways but also very different in others.

How does this change your orientation and understanding of the real-world problem? What does the calculated angle represent now?

Which picture represents, or shows, the 'truth', do you think?

◆ Assessment opportunities

- ◆ In this activity you have practised skills that are assessed using Criterion C: Communicating.

SUMMATIVE ASSESSMENT

Use these problems to apply and extend your learning in this chapter. The problems are designed so that you can evaluate your learning at different levels of achievement in Criterion D: Applying mathematics in real-life contexts.

Think again about the Statement of Inquiry for this chapter:

Finding **relationships** in closed **systems** can help us **simplify** and solve problems, using **technology** or otherwise.

How can we calculate what we cannot measure?

The purpose of this task is to find, as accurately as possible:

A **the height of your school building**

B **the length of a hanging awning (overhanging structure similar to that shown in the picture) or the length of any tilting surface or roof.**

In this assessment you will make choices and take measurements as a group, but you will perform all calculations and observations independently.

Suggested equipment

Measuring lengths: a selection of measuring devices for length such as those in the pictures.

■ Laser measurer   ■ Trundle wheel   ■ Tape measure

Measuring angles: a selection of clinometers, like these.

If you do not have a ready-made clinometer available, your group can make one with very basic classroom items. This link will show you how: **www.instructables. com/id/Basic-Clinometer-From-Classroom-Materials/**

Group work

Part A: Finding the height of the school

1 a **Estimate the height of the building before you begin.**
  b **Describe how you decided this estimate.**
  c **Sketch an appropriate diagram and record your measurements.**
  d **Make notes on your group observations, challenges and discoveries.**

Part B: Finding the length of the awning or roof

2 a **Estimate the length of the awning or roof before you begin.**
  b **Describe how you decided this estimate.**

c Sketch an appropriate diagram and record your measurements.

d Make notes on your group observations, challenges and discoveries.

Individual work

Part A: Finding the height of the school

3 a Using your measurements and a drawing, calculate the height of the school.

b How different was your answer to your original estimate?

c Describe the accuracy of the answer you have just calculated.

d Does this answer (solution) make sense in this context? Justify your answer.

Part B: Finding the length of the awning or roof

4 a Using your measurements and a drawing, calculate the length of the awning or roof.

b How different was your answer to your original estimate?

c Describe the accuracy of the answer you have just calculated.

d Does this answer (solution) make sense in this context? Justify your answer.

EXTENSION

Repeat one of your length measurements with a different measuring device and one of your angle measurements with a different clinometer.

1 State at least **one** difference and **one** similarity between the set of devices for measuring lengths and the devices for measuring angles.

2 Calculate the answer from your new set of measurements.

3 Compare both results and comment on any difference between them.

4 Suggest reasons for any differences.

5 Which measuring tools are the most accurate? Justify your answer.

6 Quantify (put a number value) on the percentage error of the less accurate solution.

Reflection

Use this table to reflect on your own learning in this chapter.		
Questions we asked	Answers we found	Any further questions now?
Factual: What is trigonometry? What is Pythagoras' theorem? What are trig ratios? How can we find any 'part' of any right-angled triangle given two values?		
Conceptual: How do geometric relationships appear in real life? Is simplification always useful? What is meant by a closed system?		
Debatable: If trigonometry can help us find unknowns, how much can we trust this in different systems and in different situations? Do trigonometric abstractions always give us the right answer in all situations? Does innovation create new mathematical relationships or help us to discover old ones in new ways?		

Approaches to Learning you used in this chapter:	Description – what new skills did you learn?	How well did you master the skills?			
		Novice	Learner	Practitioner	Expert
Communication skills					
Creative-thinking skills					
Transfer skills					
Critical-thinking skills					
Learner Profile attribute(s)	Reflect on the importance of being a thinker for your learning in this chapter.				
Thinker					

6 What is a mathematician?

O The **time**, **space** and situation we are in **justifies** the type of mathematics we use and how.

CONSIDER THESE QUESTIONS:

Factual: How has mathematics learning changed?

Conceptual: Who are we holding back? How does crowdsourcing contribute to the greater good? How does zooming in affect perimeter?

Debatable: Which is the best map projection? Is using a calculator cheating? Is mathematics really everywhere? Can animals count?

Now **share and compare** your thoughts and ideas with your partner, or with the whole class.

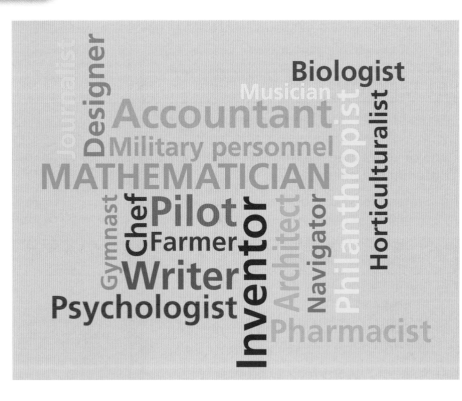

IN THIS CHAPTER, WE WILL ...

- **Find out** about important and inspiring mathematical pioneers.
- **Explore** the key concepts in a variety of mathematical areas.
- **Take action** by building up a team to attend a mathematics competition, and by participating in crowdsourcing research.

■ Affective skills
■ Information literacy skills
■ Transfer skills
■ Reflection skills

● We will reflect on these Learner Profile attributes …

● **Open-minded** – We critically appreciate our own cultures and personal histories, as well as the values and traditions of others. We seek and evaluate a range of points of view and we are willing to grow from the experience.

● **Inquirer** – We nurture curiosity, developing skills for inquiry and research. We know how to learn independently and with others. We learn with enthusiasm and sustain our love of learning throughout life.

◆ Assessment opportunities in this chapter:

◆ **Criterion B**: Investigating patterns
◆ **Criterion C**: Communicating
◆ **Criterion D**: Applying mathematics in real-life contexts

KEY WORDS

doughnut	extra terrestrial
projection	descent
pellets	

How you use mathematics depends not only on what you are doing, but on how and when. In this chapter we will revisit each related concept and see how it applies to different people, animals or things. We will encounter different professions and careers, and look at how people in these roles are also mathematicians, in their own way.

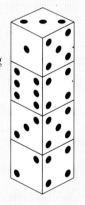

Part C

21. The number of dots on opposite faces of a regular die add to 7. Four regular dice are arranged as shown. Which of the following could be the sum of the number of dots hidden between the dice?

A   22
B   26
C   24
D   21
E   23

Source: Question from Grade 8 (M3) Canadian Mathematics contest at www.cemc.uwaterloo.ca/contests/past_contests/2017/2017Gauss8Contest.pdf

The question above was taken from a mathematics contest held by the University of Waterloo, one of the most prestigious mathematics and science universities in the world. How is this question different from one you might see on a standardized test? It is not a real-world question – why might an innovative mathematics university be interested in a student who can solve the puzzle?

How has mathematics learning changed?

COMPETITORS AS MATHEMATICIANS

While mathematics is heavily used across many industries and even in our personal lives, there is also an art behind it that is widely appreciated. Puzzles, open-ended problems, as well as rote memory skills, are subjects of intense competition throughout the world.

The Mental Calculation World Cup

Each year, Germany hosts an international competition in which competitors perform tasks like adding lists of 10-digit numbers, calculating square roots of 6-digit numbers and multiplying 8-digit numbers – all mentally and extremely quickly. Although there is a shift towards problem-solving in mathematics, as opposed to rote calculation, there is no denying that these are impressive skills.

Likewise, it is impressive that Lu Chao of China was able to enter the *Guinness Book of World Records* by reciting 67 890 digits of pi from memory. This is not necessarily a skill that will lead to breakthrough innovations, but the capacity to store so much information in human memory is to be admired. How do they do it? For most, it is not a genetic memory talent, it is a learned skill. Competitors often use the method of Loci which uses spatial visualization to remember numbers, lists of words or faces.

There is strong evidence to suggest that, with the rise of technology, our capacity to remember things is sharply decreasing. Why may this be? Why is remembering still useful in our modern age? Discuss with a partner.

ⓘ Change: A variation in size, amount or behaviour.

Spatial visualization

■ ATL

- Information literacy skills: Use memory techniques to develop long-term memory
- Affective skills: Managing state of mind

Read about spatial visualization here: **https://www.livescience.com/50134-pi-day-memory-experts.html**. Now try to use the techniques described in the article to remember something useful, such as:

- the phone numbers that are stored in your phone
- recipes
- your family tree
- former leaders of your country.

Beyond memory feats, however, there are many competitions on national, regional and international levels that involve solving complex problems within a given time frame. The Philippines hold 26 national mathematics competitions per year! Since 1959, pre-college student teams of up to six students have attended the International Mathematical Olympiad. This two-day, six-question competition includes problems that appear deceptively simple but which require a 'certain level of ingenuity'. It is currently funded by Google.

The Mathematical Contest in Modeling in the USA has a strong focus on teamwork, communication and justification of results, and there is also an Interdisciplinary Contest in Modeling that takes place concurrently, applying mathematics to other disciplines. There is a China Girls Mathematical Olympiad which, despite its name, invites girls from other countries to participate, and the Mathematical Kangaroo, which includes grades 1–12. Search for a list of international mathematics competitions. What is available in your country? Is your school involved?

$X + Y$

ATL

- Reflection skills: Consider content
- Information literacy skills: Create references and citations, use footnotes/endnotes and construct a bibliography according to recognized conventions

The movie $X + Y$ tells the story of a boy who competes in the International Mathematical Olympiad. Watch the film and write a brief review that addresses the following questions:

1 How does being a part of the International Mathematics Olympiad change Nathan's life? The lives of those around him?

2 How else can **change** be seen in this movie?

3 What did you learn from the film?

4 What questions do you have about it?

Where applicable, include quotes from the movie that especially stood out to you. Remember to use your school's bibliography conventions for citing footnotes.

As mathematicians immerse themselves more deeply into their subject, and develop it further or apply it directly, they may be considered for the most prestigious honour in mathematics – the Fields Medal, often described as the 'mathematician's Nobel Prize'. It is awarded for outstanding contributions in mathematics from young scientists. It aims to present the work of each Fields medallist and its impact, to explore the potential for future directions and areas of influence, to provide inspiration to the next generations of mathematicians and scientists, as well as to present the medallist to a broader public.

■ The Fields Medal for young mathematicians

DISCUSS

An intelligence quotient (IQ) is a number designed to measure human intelligence. It is assessed through a series of standardized tests which aim to determine a person's capacity for understanding, as opposed to the amount of knowledge they currently possess. In the 20th and 21st centuries, average IQ seems to have increased – does this mean the world's population is getting smarter? In a person's lifetime, their IQ may increase or decrease – can you think of reasons for these changes?

EXTENSION

A number of IQ tests are available. Try a few of them and see if your score changes or is relatively consistent from test to test. Why might this be? Is this a good way to measure intelligence?

How has mathematics education changed?

Why is there a shift to problem-based, 'discovery' learning? Is there still a place for memorization in mathematics? Would a fundamentalist or a discoverist make a better competitor? Read the two articles below before writing your response.

Fundamentalists: http://nationalpost.com/news/canada/math-wars-rote-memorization-plays-crucial-role-in-teaching-students-how-to-solve-complex-calculations-study-says

Discoverists: www.thestar.com/news/canada/2016/09/03/no-teaching-math-the-old-fashioned-way-wont-work-wells.html

❗ Take action

❗ Find a mathematics competition that appeals to you. Help organize a mathletes team for your year or younger. Hold weekly meetings to practise past questions and host a district or network competition to prepare for your target competition. Start a crowdfunding page to sponsor a competitor who can't afford to go.

Can animals count?

ANIMALS AS MATHEMATICIANS

Assigning a number to 'nothing' can be a counterintuitive thing. Like infinity, many of us think of zero as a symbol rather than a number. However, isn't every number used today simply a symbol of a certain quantity, which we have generally accepted as 'equivalent' to that quantity? While numbers as we know them are a human-made system of symbology, research is revealing that some animals are able to differentiate between quantities as well – though they may not assign symbols or names to those quantities as we do.

Salamanders, three-day old chicks and honeybees are among animals that are able to perform basic arithmetic. Chimpanzees, when presented with two pairs of bowls containing various chocolate pieces, are able to **add** the contents of each pair of bowls, **compare** them and select the larger sum! Desert ants, rather than retracing their steps (which they can't do as there are no landmarks or scent trails in the desert), calculate the most direct route home. They count their steps to calculate distance, and determine their direction using trigonometry based on the angle between their path and the Sun's position. As the Sun moves across the sky throughout the day, their calculations are constantly updated. Search Animals that count: how numeracy evolved for an interesting video on animal recognition of numbers.

ⓘ Equivalence: The state of being identically equal or interchangeable, applied to statements, quantities or expressions.

Animals that count

■ ATL

- Information literacy skills: Access information to be informed and to inform others; Make connections between various sources of information

Select one of the animals listed, or come up with another, and provide a brief presentation to inform others of the animal's understanding of the 'numbers' equivalent. Try to include multimedia in your presentation to really convince your audience!

- baby chicks
- honeybees
- gorillas
- rhesus monkeys
- capuchin
- squirrel monkeys
- lemurs
- dolphins
- elephants
- birds
- salamanders
- fish

EXTENSION
The odd ball

There are 12 billiard balls, all the same size, shape and colour. All weigh exactly the same, except the odd ball which is very slightly different in weight. It could be heavier or lighter, but is not noticeably so in the hand.

Using a beam balance, you have three opportunities to weigh the balls and identify the odd ball.

Stuck? Search for weighing pool balls puzzle and see which website communicates the solution best!

MEET A MATHEMATICIAN: MARYAM MIRZAKHANI (1977–2017)

Learner Profile: Inquirer, Open-minded

At Harvard University, where Mirzakhani earned her PhD under scholarship, she was 'distinguished by her determination and relentless questioning'.

'I was struck by her intense curiosity and drive.'

Professor Curtis McMullen

The first woman and the first ever Iranian to win a Fields Medal, and a working mum, Maryam Mirzakhani possessed overwhelming curiosity, insightfulness and imagination.

Most of her work revolved around counting the number of possible pathways along the surface of a doughnut. Remember what 'surface' means? Think back to the nets you studied in Chapter 3. What would a spherical (sphere-shaped) net look like? What would it look like with a hole in the middle? Two holes? Infinite holes? Maryam was able to connect ideas from different areas of mathematics to make sense of such unimaginable shapes and to solve problems about them. She explained this once by referring to billiard tables of every possible shape.

Listen to the *More or Less* podcast about Maryam Mirzakhani (Maryam Mirzakhani – A Genius of Maths) and answer these questions:

1 What is meant by counting loops on a doughnut?

2 What real-world applications does her research apply to?

3 What is the Mirzakhani Society?

ℹ️ **Generalization**: A general statement made on the basis of specific examples.

FARMERS AS MATHEMATICIANS

Many cultures associate particular foods with different seasons – apples and pumpkins with the autumn, berries with high summer, and corn in late summer. This is no coincidence. Planting and harvesting of crops have always been linked to seasonal observations – and the prediction of seasons is the result of the same astronomical calculations that led to modern-day calendars. Farmers have been noting weather patterns, soil fertility, and Sun and Moon phases for centuries, generalizing their findings to predict the sorts of crops that would be in abundance in the coming months and years.

Farmers turn to **almanacs** for information like tide tables, weather forecasts and celestial statistics, and by generalizing and comparing past experiences, they can come up with planting dates. Farmers are often mathematicians and scientists without even knowing it – testing soils, planting locations and watering patterns for example.

Even before the first crops were farmed, people who were involved in animal husbandry were generalizing patterns in birthing, feeding and breeding of animals for agricultural purposes. In understanding these patterns and generalizing them, farmers were able to raise livestock for milk, eggs and meat.

Modern farmers make similar observations but about very detailed information. They are looking for patterns to generalize statements about pesticides, crop rotation, dosages of chemicals, percentages of nutrients, and so on. Many of these ideas are still in development and statements may change or become more streamlined over the years. In Japan, the calendar has been generalized into 72 distinct micro-seasons: **www.nippon.com/en/features/h00124/**.

Without farming, the need for keeping accurate calendars may not have arisen as early or as quickly.

THINK-PAIR-SHARE

Use this link **www.nippon.com/en/features/h00124/** or search calendar tricks and play a few on each other! Share your favourites with the class.

Which is the best map projection?

ⓘ **Justification**: Valid reasons or evidence used to support a statement.

CARTOGRAPHERS AS MATHEMATICIANS

If we study patterns in the Cartesian plane, we find increased purpose and meaning in mathematics. We see direct links to algebra and to geometry, often forgetting that these disciplines were first created for cartography purposes. Recall the story of Descartes and the fly on the ceiling in *Mathematics for the IB MYP 2*. Mapping out the location of a fly on a two-dimensional ceiling using a grid is fairly straightforward, but designing a grid system or map to represent three dimensions is an entirely different challenge.

ACTIVITY: Can maps be wrong?

■ ATL

■ Information literacy: Evaluate and select information sources and digital tools based on their appropriateness to specific tasks

1 **Watch the six-minute video 'Why all world maps are wrong':** https://youtu.be/kIID5FDi2JQ

2 **Research the following map projections and justify their use in the real world. What do they preserve? (Consider distances, areas and angles, for example.) What do they omit? Who might want to use each map projection?**
   - Mercator
   - Robinson
   - Dymaxion
   - Winkel-Triple
   - Goode-Homolosine
   - Hobo-Dyer
   - Plate Caree
   - Gall-Peters

3 **Now that you know about map projections, examine a full-size version of the cartoon at https://xkcd.com/977/. Explain the reason the author of this cartoon artwork chose those personality traits.**

◆ Assessment opportunities

◆ In this activity you have practised skills that are assessed using Criterion C: Communicating and Criterion D: Applying mathematics in real-life contexts.

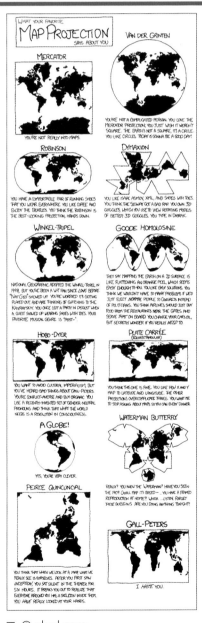

■ © xkcd.com

How does crowdsourcing contribute to the greater good?

ⓘ **Measurement** A method of determining quantity, capacity or dimension using a defined unit.

SCIENTISTS AS MATHEMATICIANS

As we have learned from past statistics units, the more data we have, the more accurate our predictions can be. **Citizen Science** is a scientific method that uses crowdsourcing for data collection. As scientists search for big data, they outsource data collection to the general public, often via the internet. Crowdsourcing also includes collecting information, such as Wikipedia has done, or delegating small tasks to a huge number of people in order to finish a project much sooner. Let us focus on data collection by looking at some examples.

The Great Nurdle Hunt

www.nurdlehunt.org.uk/

Many plastic products use tiny plastic pellets – nurdles – in their production. Billions of nurdles are used each year and alarming quantities end up in the sea. They attract toxins and are small enough to be ingested by marine animals and birds. The Great Nurdle Hunt website 'collects evidence to show the local plastics industry the extent of the pollution'. Participants search local beaches and report the location and quantity of nurdles found.

Flowers for Bees

www.zooniverse.org/projects/eileenfranklin/count-flowers-for-bees

When volunteers identify and count the flowers in the images provided by this website, it allows researchers to find locations that are good for pollinators like bees, and to highlight the places where bees need help. The

■ How many nurdles are at your nearest beach?

bee population has declined in recent years and many species are threatened with extinction. Without their help in pollinating, our crops and our very ecosystem are in danger.

Galaxy Zoo

www.zooniverse.org/projects/zookeeper/galaxy-zoo/

Galaxy zoo contains a wealth of images of galaxies in the night sky and asks the public to help in classifying them to aid the research of galaxy evolution.

Protein Foldit

https://fold.it/

The objective of Foldit is to fold the structures of selected proteins as perfectly as possible, using tools provided in the game. The highest scoring solutions are analysed by researchers, who determine whether or not there is a native structural configuration (native state) that can be applied to relevant proteins in the real world. Scientists can then use these solutions to target and eradicate diseases, and create biological innovations.

SETI@home

https://setiathome.berkeley.edu/

The Search for Extra Terrestrial Intelligence transmits radio frequencies from their Allen Telescopic Array and streams them. Participants are asked to search for signs of unusual activity, in the hope that humans may find things the automatic system might miss.

Crowdsourcing projects are often time-sensitive and have closing dates to ensure they have a final measurement. Some also ask for use of people's computer's spare processing power, for example in prime number or GIMPS hunting.

We can use big data to track ships, flights, whales, sharks and hurricanes for example in real time, using apps like flightradar24 and shipwatch.

#stormireland

James B @weatherman   25s
This is cool! Tracking **#stormireland** on this link:
https://earth.nullschool.net/

Village voice @voiceofthevillage   2m
Storm damage round here frightening. Trees down, chimney collapsed, roof off. Stay safe everyone. **#stormireland #staysafe**
pic.twitter.com/fg458ghJn

Belle B @bluebell   3m
Been awake half the night listening to the winds! #stormireland

Jill O @mrsoneill   5m
Just arrived in Dublin. Hats off to the plane crew. Some landing. **#stormireland**

 Home   @ Connect   # Discover   Me

■ Even hashtags are a form of crowdsourcing. While not always credible, they can help make real-time information possible.

Facial recognition software helped the team isolate individual selfies, and using Instagram's API (application programming interface), researchers created Selfiexploratory, an interactive project that displays several data points Monvich's team analysed: age, sex and the degree of a selfie snapper's head tilt, among them. The results show that selfie aficionados in Bangkok skew younger than those in, say, Berlin. People in Moscow smile far less than their happier, more expressive counterparts in São Paulo.

Source: www.theguardian.com/media/2014/feb/20/how-people-selfie-around-the-world

! Take action

! Choose a credible crowdsourcing site and contribute to its research. Create a brochure detailing the site's research, intended use of data and how you contributed, and use it to inspire others to join in.

Is using a calculator cheating?

GRAPHIC DISPLAY CALCULATORS AS MATHEMATICIANS

Early in your mathematics studies, you probably learned to solve simple addition or multiplication problems without the use of a calculator. These are important skills as they can help you determine in a moment whether a solution makes sense (for example, two packs of gum costing 50 euro) or to save time when performing other calculations. These are not so time-consuming to calculate mentally, and are often performed faster **without** a calculator.

Try it! Ask two friends a single-digit multiplication question at the same time. One must answer back mentally, and the other must enter it into their calculator. Who was fastest?

After nearly a year in M3, you have learned to create equations to model real-life problems, and to draw graphs that do the same. A calculator cannot choose techniques or develop equations for you – but what if it could save time in **solving** or **graphing** them for you?

You must be able to graph linear equations and solve algebraic equations without a calculator. It is important to understand **how** these calculations are performed, but suppose your real-life situation includes 'messy' numbers? Large numbers, with many decimals? There is so much room for human error, and the end goal is to solve a problem accurately, not to spend three days on the calculations.

ACTIVITY: The rental car model

■ ATL

■ Critical-thinking skills: Gather and organize relevant information to formulate an argument

Recall from Chapter 2, the car rental agency that charged £24.99 to rent a standard sized car for one day, plus £0.25 per km driven:

$y = 0.25x + 24.99$ where x represents the number of km driven, and y represents the total cost

Locate the 'y=' button on your calculator. This takes you to a screen where you can enter all kinds of relationships. Type the rest of the equation. The 'x' can be found next to the green 'ALPHA' key.

Now select 'TABLE'. Notice 'TABLE' is in blue, above the graph key. So, we will need to press the blue '2nd' key before selecting 'TABLE' – once the blue key is pressed we can use any process on the calculator that is written in blue.

Write a sentence about what the values in the table represent. What is the difference between each y-value? Why is that?

Now press the 'GRAPH' button. Why is it blank? (Hint: what is your y-intercept?) Press the 'window' key.

ⓘ **Model**: Depiction of real-life event using an expression, equation or graph.

What do you suppose these values mean? What will you need to change them to in order to make your graph visible on the screen? Write down the values you choose to use to better analyse this graph.

Suppose a competing company offers to rent their standard cars at a flat rate of £35, with unlimited mileage (so that you can drive as far as you like with no extra charge). Let's add this equation to our graphing screen. Go to '*y*=' and for y_2 enter 35. (Why is there no *x*?) Ensure your window settings are adequate and press 'GRAPH' again. Explain how the two rental companies compare. Now go to the table – can you compare more accurately with this information?

Investigate what the 'TRACE' button does. Notice the values at the top and bottom of the screen.

Now select 'CALC' (in blue above 'TRACE'). Investigate what '#1: value' and '#5: intersect' do. How might these be helpful in deciding on a car rental company?

Assessment opportunities

◆ In this activity you have practised skills that can be assessed using Criterion B: Investigating patterns and Criterion D: Applying mathematics in real-life contexts.

PILOTS AS MATHEMATICIANS

Aircraft pilots have to make complex calculations quickly. In regularly making many of these calculations they soon recognized patterns in the results and devised shortcuts. None of these shortcuts are actually mathematical theorems as they don't use formulae – they use quicker techniques based on the observed patterns. But make no mistake, the calculations are still performed mathematically every time!

When to begin descent

Aircraft cockpit displays give pilots readings of their altitude and the altitude of the runway they will be landing on. They use the difference in these numbers, in feet, to calculate how soon they must begin their descent towards the ground.

Patterns: Sets of numbers or objects that follow a specific order or rule.

- 40 000 ft → begin 120 miles out
- 60 000 ft → begin 180 miles out
- 100 000 ft → begin 300 miles out

Spot the pattern? Pilots simply omit the last three digits of the altitude difference and multiply that by 3.

How fast to descend

The speed of descent, together with the distance, ensures the plane is descending at the correct angle. Pilots start with their ground speed in miles per hour and convert to their speed of descent in feet per minute.

- 500 mph → 2500 fpm
- 700 mph → 3500 fpm
- 200 mph → 1000 fpm

Spot the pattern? They have taken half the ground speed number and added a zero.

How much fuel to order

In the USA, pilots read the required fuel amount from the display in pounds (lbs) but must convert this to gallons (gal) in order to tell the fuel operative what they need. Here are a few examples of these conversions.

- 3000 lbs = 450 gal
- 5000 lbs = 750 gal
- 200 lbs = 30 gal

Spot the pattern? Take away the final digit, and add 50%.

Pilots are required to pass a written geometry and algebra exam before they even begin to fly. Test pilots, who assist in the design and construction of aircraft, need an even more thorough comprehension of geometry, physics and trigonometry in order to understand how weights and balances, wind, and wing shapes can affect flight.

How does zooming in affect perimeter?

Quantity: An amount or number.

FINE ARTISTS AS MATHEMATICIANS

Imagine you were tasked with determining the length of Britain's coastline. It would take far too long to physically walk around the entire coast with a string and, even then, your accuracy would not be perfect. Perhaps you chose to use a pole that is 200 km long to save time? (See image below.) You end up with a total coastal length of 2400 km, but are sceptical of its accuracy. You try a 100 km pole instead, which gives you a total coastal length of 2800 km, and finally a 50 km pole – the coastal length turns out to be 3400 km. In using smaller and smaller units of measurement, we might expect these numbers to increase at a slower rate until finally they hover around one number – this is called **converging**. Instead, the more we 'zoom in', the greater the number of kinks and wiggles in the coastline we must take into account, which makes the total length larger and larger. As the units of measurement are made smaller, the quantity of length becomes infinitely larger.

Mathematics and art

ATL

- Transfer skills: Make connections between subject groups and disciplines

Describe how mathematics is seen in the following forms of art:
- indigenous art
- string artists
- classical art (golden ratio)
- negative space.

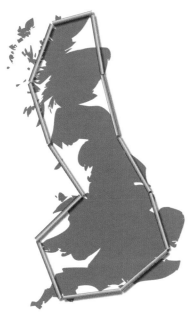

Unit = 200 km
Length = 2400 km (approx.)

Unit = 100 km
Length = 2800 km (approx.)

Unit = 50 km
Length = 3400 km (approx.)

This is a phenomenon known as the **Richardson Effect**, after Lewis Fry Richardson.

ACTIVITY: Frozen fractals all around

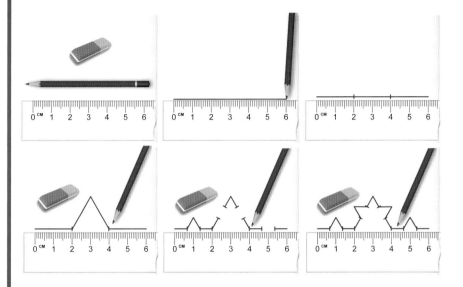

1 Using a pencil, draw a straight line segment, 6 cm long.
2 Erase the middle 2 cm and draw two 2-cm lines that meet at a point.
3 You now have four 2-cm-long line segments, or a total of 8 cm. Erase the middle 1 cm in each line segment and draw 1 cm lines meeting at a point, as before.
4 You now have sixteen 1 cm-long line segments, or a total of 16 cm.
5 Repeat twice more in a similar manner, estimating line lengths if needed. What is this beginning to look like? What is happening to your perimeter?

EXTENSION

Create your own fractal snowflake.

1 Begin with a shape – any shape! Here's an example.

2 Create five more copies and rotate them about the same centre point.

3 Take this new image and create five more, rotating them about the same centre point.

4 Repeat as desired.

For more fun with fractals, search for fractal videos. There are some hypnotizing ones!

DISCUSS

Where did we see a similar pattern in Chapter 5?

◆ Assessment opportunities

◆ In this activity you have practised skills that can be assessed using Criterion B: Investigating patterns.

Who are we holding back?

PIONEERS AS MATHEMATICIANS

A number of people who have excelled in mathematics did so despite obstacles that others never had to face. They have had to fight against discriminatory representations of themselves in order to contribute to the same world that chose to show prejudice against them. In doing so, they have helped to dismiss stereotypes.

> **ⓘ** **Representation**: The manner in which something is presented.

Hedy Lamarr was an actress during Hollywood's 'Golden Age' whose beauty was always emphasized and who was given very few lines in her acting roles. At a time when women were just entering the workforce and fighting to be taken seriously, Howard Hughes – American business man and one of the most famous pilots in history – was consulting Lamarr regarding improvements in aviation; he called her a genius. During the Second World War, the starlet invented a secret communication system for radio-guided torpedos, which became the foundation for GPS, wifi and bluetooth technology.

Benjamin Banneker – self-educated mathematician, astronomer, compiler of almanacs and writer – surveyed the territory for construction of America's capital, Washington DC. He sent one of the almanacs with his astronomical calculations to then president Thomas Jefferson, who 'considered it as a document to which (his) whole colour had a right for their justification against the doubts which have been entertained of them'.

Benjamin Banneker

Black Heritage USA 15c

Mary Somerville was heavily discouraged from studying mathematics – first by her father who believed women's minds could not handle too much academic effort without causing severe health damage, then by her first husband who did not believe women should have any intellectual pursuits. She studied in secret for the majority of her life, until her second husband (who was proud of his wife's intellect) presented her research papers to the Royal Society as women were not permitted to attend. She is most noted for writing the first English geography text, *Physical Geography*.

Danica McKellar is a well-known Hollywood actress who, after earning her Mathematics degree from UCLA, combined her fame and knowledge to break the stereotype of the 'mathematics nerd', and wrote several books aimed at teaching and inspiring girls in mathematics. She is also the co-author of 'a ground-breaking mathematical physics theorem', the Chayes-McKellar-Winn Theorem.

Alan Turing, English mathematician, logician and cryptographer, broke the Nazi Enigma code, contributing to their defeat in the Second World War and leading to the creation of the computer. He also invented the theoretical idea of software. Turing was 'punished for homosexuality' with a series of hormone injections. Friends say this was a visibly traumatic experience, but he fought back by continuing to work as though nothing had happened. Because homosexuality was considered a security risk at that time, Turing was not permitted to continue his secret work. Nobody knew just how much he had contributed to the war effort until 60 years after his death.

How do computers simplify our lives?

DISCUSS

Have personal computers and the progression of technology made our lives simpler or more complicated? Or both? In what ways?

PROGRAMMERS AS MATHEMATICIANS

If you watch any movie made before the year 2000, you may think how abruptly the plot could have concluded if mobile phones and internet access had been available back then. So much of life has been simplified through the use of computers. In the past, travellers needed map books for every destination – now we have GPS. People wrote letters on paper to keep in touch with overseas friends and family – now we skype. We used to pay our bills in person at banks, conduct our research using books at the library, and wait entire weeks between episodes of our favourite television programmes!

Computers have an incredible ability to take very simple inputs and produce extremely complex outputs. A complex computer-generated image, for instance, is really just a collection of tiny dots called pixels. The first ever fully automatic digital computer used a binary system to function – just like the one in the clock activity you tried in Chapter 6 of *Mathematics for the IB MYP 2*. Simple 1s and 0s were transformed into complex calculations, words, commands and images. At its core, programming is very simple, but programs can do very complex things.

Programmers use code to write software, programs or apps for computers. Using a fairly simple language, they are able to tell the computer what to do. Their biggest challenge is often taking a complex request and breaking it down into very simple tasks. To assist them in this challenge, programmers are avid users of flowcharts and algorithms, which they usually employ before ever writing a word of code. Recall from Chapter 3, that algorithms present the processes that are used to solve problems – or at least bring them to their simplest form. Something like adding two numbers sounds pretty simple, but it actually takes **six** steps for a computer to perform the action!

```
Step 1:  Start
Step 2:  Declare variables num1,
         num2 and sum.
Step 3:  Read values num1 and num2.
Step 4:  Add num1 and num2 and assign
         the result to sum.
         sum←num1+num2
Step 5:  Display sum
Step 6:  Stop
```

Source: **https://www.programiz.com/article/algorithm-programming**

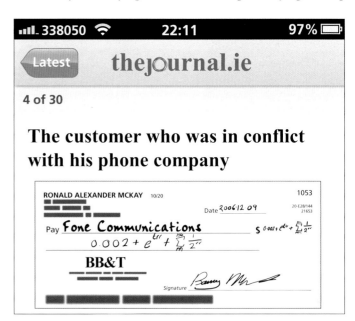

■ Mathematics can be used to simplify or complicate things!

ARCHITECTS AND DESIGNERS AS MATHEMATICIANS

Architects must combine a unique aesthetic with practical and safe functionality when they design buildings and other structures. A caring architect will also take extra effort to minimize the environmental footprint their structure will leave. Many modern architects incorporate complex geometrical figures into their style, but this has actually been a trend since antiquity. Egyptian and Mayan pyramids, European archways and Japanese pagodas are just some of the geometrically fascinating structures that have been created in the distant past. As we learn more about technology, it enables us to create more and more complex structures.

■ The architect Zaha Hadid used complex geometric figures to create her unique and artistic structures. The challenge for every architect, of course, is to make these structures beautiful **and** functional, and both are achieved through the use of mathematics.

A great deal of science has always gone into the construction of concert halls to ensure they distribute sound properly, as opposed to muting it or producing echos. A field of science called **acoustics** specializes in the use of space to carry sound in an ideal way. The secret of using acoustics in a concert hall lies not in the construction of the perfect panel and then replicating it all around the hall, rather *different* panels must be arranged in relation to one another in such a way that the sound waves bounce from them in balance. This has always required a huge effort and years of research prior to construction, but the newly built Elbphilharmonie in Hamburg, with its 10 000 panels, is especially complicated. Each panel contains an average of 100 cells of varying depths and dimensions to distribute sound. As the lead designer, Benjamin Koren, stated, 'it would be insane to do all that by hand'. Instead, he used 'parametric design' – a design technique which takes certain parameters from the designer (in this case, acoustic and artistic considerations) and creates a million unique cells that meet the specifications. He said, 'I have 100 percent control over setting up the algorithm, and then I have no more control.'

Is mathematics really everywhere?

CREATIVES AS MATHEMATICIANS

Music making

J.S. Bach's music followed very intricate and symmetrical structures. When graphing the pitch frequencies of his canons, for example, we see translated functions every time.

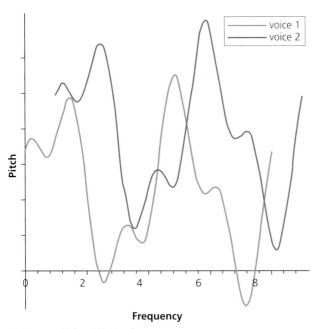

■ Canon 5 by J.S. Bach

Voice 2 begins one measure after voice 1, and is a perfect fifth higher. Imagine singing 'Row, row, row your boat' and someone joins in after the first four beats and at a different pitch, but it all harmonizes together beautifully!

In his book *Chronicles*, Bob Dylan refers to a playing style he learned 'based on an odd instead of even numbered system'. It is suggested that in Dylan's early days his music came from his own inspiration but that, like most artists, he ran into writer's block and looked to mathematics to help him strategize his writing. It

comprised four elements: (1) a certain approach to rhythm and (2) melodic cells, (3) based on more or less esoteric considerations of the power of numbers, (4) which, taken together, make up a formulaic system.

Philip Glass writes opera connected to science. He studied music at the famous Julliard School of Music, and studied mathematics at the University of Chicago. He connects the aesthetics of music to that of mathematics:

> 'The beauty of mathematics is something that mathematicians talk about all the time, and the elegance of a mathematical theorem is almost as good as its proof,' he said. 'Not only is it true, but it's elegant. So you get into almost aesthetic questions.'

Poetry – rhythmic systems

The very structure of poetry begs a mathematical formula to keep it in balance. Take this mutilation of Robert Louis Stevenson's 'To My Mother', which opens *A Child's Garden of Verses*:

> 'You too, my mother, read my rhymes
>
> For love of unforgotten times,
>
> And you may chance to hear once more
>
> The pitter patter of little feet running along the floor.'

Something is wrong with that last line. It does not follow the pattern and symmetry of the rest – and we anticipate it. It is a human expectation for this to end in four beats. Thankfully, Stevenson was well aware of a mathematical rule which poets refer to as 'metre', and followed it:

> 'You too, my mother, read my rhymes
>
> For love of unforgotten times,
>
> And you may chance to hear once more
>
> The little feet along the floor.'

	Humans	Non-humans	Total
For Alice	5	12	
Against Alice	10	15	
Total			

■ A Carroll diagram showing classification of Wonderland residents.

Mathematics in writing is not limited to the form of the work. Some writers embed mathematics directly in their text.

The Reverend Charles Lutwidge Dodgson was a mathematician at Christ Church, Oxford. He is known in the world of mathematics for his creation of the **Carroll diagram**, also known as a two-way table. He was better known in the world of literature for his famous *Alice in Wonderland* books – written under the pen name of Lewis Carroll. Interwoven among the well-known stories are all kinds of mathematical references – some abstract and others more focused.

Dodgson was a traditional mathematician during a time when abstract and theoretical mathematics was rapidly growing. It is believed that many of his references to 'nonsense' were in fact references to his view of what mathematics was becoming. The Cheshire cat who disappears leaving only his smiling mouth is said to represent the declining standards of rigour. In the story, Alice moves from a rational world to one where numbers behave unpredictably. She tries to remember her multiplication, but the numbers are no longer in the base-10 system. The character of the hookah-smoking caterpillar is Arabian-based – just like the word 'algebra', which translates to 'restoration and reduction', mirroring Alice's constant struggle whenever she eats or drinks. Just before parting ways, the caterpillar tells Alice to 'keep your temper' which Alice interprets as 'don't get angry'. It is an odd thing to say as she hadn't appeared so. The word 'temper' also means to mingle in due proportion – so he may well be telling her that to survive Wonderland she can be big or small but must maintain her proportions – much like the isometric transformations we studied previously in *Mathematics for the IB MYP 2*, Chapter 5. This is another reference to his love of traditional mathematics.

Simon Singh is another mathematician and author who incorporates mathematics directly into his writing. Watch an episode of *The Simpsons* or *Futurama* and look out for one of his many 'freeze-frame gags' – visual quips that go by so quickly they would only be noticed when the show is paused. Famous formulae and odd terminology grace the spines of books, the titles of public gathering places and TV screens. There are also less subtle moments that are still only noticed by those who keep up with their mathematics. For instance, in an effort to quieten down and gain attention from his colleagues at a conference, Simpsons' character Professor Fink asks for order with little success – then announces that pi is exactly three and all eyes fall silently on him. To hear Singh discuss this himself, search **Simon Singh, the Simpsons and their mathematical secrets**.

Read this fascinating article about the Netherland's amazing **natural** solution to feeding a huge population with a lack of agricultural land:

www.nationalgeographic.com/magazine/2017/
09/holland-agriculture-sustainable-farming/?utm_
source=NatGeocom&utm_medium=Email&utm_
content=wildscience_20170916&utm_campaign=
Content&utm_rd=961341852

Describe at least five very interesting facts you noticed in the infographics of this article.

How are infographics used in journalism? Would you say journalism is a system whose credibility relies on mathematics? Why or why not?

■ Even naval enthusiasts find ways to relate mathematics to their boating passion!

SUMMATIVE ASSESSMENT: #Actuallivingscientist

Use these problems to apply and extend your learning in this chapter. The problems are designed so that you can evaluate your learning at different levels of achievement in Criterion C: Communicating.

In February 2017, a scientist named David Steen gave the following tweet:

David Steen Ph.D. ✅
@AlongsideWild

Most Americans can't name a living scientist (it's true, look it up!); no wonder they don't fully appreciate what we do. So ... Hi, I'm Dave.

3:29 PM - Feb 2, 2017

♡ 131   ⎘ 948   ♡ 3,113

David Steen Ph.D. ✅
@AlongsideWild

I'm Dave & an #ActualLivingScientist. I want to know how animals use and persist on landscapes so we can live alongside them at the same time.

10:54 AM - Feb 5, 2017

♡ 17   ⎘ 144   ♡ 621

An entire movement followed, and scientists and mathematicians around the world began introducing themselves. We have met many mathematicians since – some are living but many are from the past. Your task is to introduce a living mathematician who has not been featured in the *Mathematics for the IB MYP* books. Select a Learner Profile attribute that best describes them, and create a one-page fast facts paper about them. Be sure to include their photograph, major achievements, and any interesting information about their origins, obstacles they overcame or other fun facts you discover. Present your mathematician to the class and then add them to your classroom's 'Wall of #Actuallivingmathematicians'.

Reflection

Questions we asked	Answers we found	Any further questions now?			
Factual: How has mathematics learning changed?					
Conceptual: Who are we holding back? How does crowdsourcing contribute to the greater good? How does zooming in affect perimeter?					
Debatable: Which is the best map projection? Is using a calculator cheating? Is mathematics really everywhere? Can animals count?					
Approaches to Learning you used in this chapter:	Description – what new skills did you learn?	How well did you master the skills?			
		Novice	Learner	Practitioner	Expert
Affective skills					
Information literacy skills					
Transfer skills					
Reflection skills					
Learner Profile attribute(s)	Reflect on the importance of being open-minded and an inquirer for your learning in this chapter.				
Open-minded					
Inquirer					

Glossary

Contiguous sharing a common border, or adjacent to one another

Cosine the ratio, or relationship, between the adjacent and the hypotenuse expressed as a quotient

Curvilinear grid a grid consisting of curved lines rather than straight lines

Direct variation when two variables are directly related

Fractal a curve or geometrical figure, each part of which has the same statistical character as the whole

Fundamental counting principle a method for figuring out the number of outcomes in a probability problem

Gradient the rate of change, or incline, of a line

Indefinite hyperbolic numerals words to describe large numbers, such as million, billion and trillion

Indirect variation a relationship between two variables in which the product is a constant

Lemniscate the infinity symbol

Mass measurement measuring or putting a quantitative value for mass

Network a group or system of interconnected things, including people

Proof a deductive or reasoned argument for a mathematical statement

Pythagoras' theorem the theorem which relates the length of sides of a right-angled triangle

Quotient a fraction or a ratio

Set a collection or group based on some common condition

Simultaneous at the same point in time and space

Sine the ratio, or relationship, between the opposite and the hypotenuse expressed as a quotient

Standard form an equation that takes the form Ax + By = C, where x and y are variables and A, B and C are integers

Subsets a set fully contained, or within, another set

Tangent the ratio, or relationship, between the opposite and the adjacent expressed as a quotient, or a line which touches but does not cross within a circle

Acknowledgements

The publishers would like to thank the following for permission to reproduce copyright material.

Photo credits

t = top, *l* = left, *c* = centre, *r* = right, *b* = bottom

p.3 *tl* © Aleksandar Todorovic/stock.adobe.com, *tr* © Tursunbaev Ruslan/Shutterstock.com; **p.4** *tl* © NASA Photo/Alamy Stock Photo, *cr* © aitormmfoto/stock.adobe.com; **p.6** *tl* © swisshippo/ stock.adobe.com; **p.10** *tr* © Toru Yamanaka/AFT/Getty Images; **p.11** *cr* © antlia/stock.adobe.com; **p.12** *tr* © s72677466/stock.adobe.com, *cr* © Joseph Ostermeier, President, CEO, Infinity Records USA; **p.16** *tl* © https://commons.wikimedia.org/wiki/File:%E4%B9%9D%E7%AB%A0%E7% AE%97%E8%A1%93%E7%B4%B0%E8%8D%89%E5%9C%96%E8%AA%AA.jpg, https:// creativecommons.org/licenses/by-sa/3.0/; **p.17** *br* © Royal Astronomical Society/Science Photo Library; **p.21** *tl* © porcomanzi/stock.adobe.com, *bl* © Theastock/stock.adobe.com; **p.23** *br* © Rita Bateson; **p.28** *br* © Festival of the Spoken Nerd photograph by Mihaela Bodlovic; **p.31** *t* © Brian Jackson/stock.adobe.com; **p.32** *tr* © Elenathewise/stock.adobe.com; **p.48** *cr* © Alison Wright/ Danita Delimont/Alamy Stock Photo; **p.54** *tr* © Hellen/stock.adobe.com; **p.55** *tl* © shaineast/ stock.adobe.com, *tr* © djahan/stock.adobe.com; **p.56** *bl* © imageoptimist/stock.adobe.com, *tr* © Antonio Abrignani/123RF, *cl* © Fine Art Images/Heritage Image Partnership Ltd/Alamy Stock Photo, *cr* © Weinstein Company/Everett Collection Inc/Alamy Stock Photo; **p.57** *tl* © Pixeljoy/ Shutterstock.com, *br* © Harismoyo/stock.adobe.com; **p.58** *tl* © cegli/stock.adobe.com, *bl* © Nic Hamilton Photographic/Alamy Stock Photo, *tr* © goodwin_x/stock.adobe.com, *br* © irazzers/ stock.adobe.com; **p.59** *tr* © Sebastian Derungs/AFP/Getty Images, *br* © Gorilla/stock.adobe.com; **p.61** *cl* © Copyright by the International School of Hamburg, *tr* © https://commons.wikimedia. org/wiki/Category:Diophantus, https://creativecommons.org/licenses/by-sa/3.0/, *br* © VanderWolf Images/stock.adobe.com; **p.63** *tr* © Hannah Fry; **p.64** *tr* © Art Collection 4/Alamy Stock Photo; **p.66** *cr* © lindama/stock.adobe.com; **p.69** *tl* © vichie81/stock.adobe.com; **p.72** *tl* © Drawing by Manu Cornet, *tr* © Emotion shapes the diffusion of moralized content in social networks, William J. Brady, PNAS,2017.114(28) 7313-731, *br* © Anita Ponne/stock.adobe.com; **p.73** © International Baccalaureate®; **p.74** *tl* © Marek/stock.adobe.com; **p.78** *cl* © Photo by Adam Morganstern; **p.80** *tl* © OpenStreetMap contributors, CC-BA-SA 2.0, *tr* © Loci by Andrew Spitz, *br* © Huseyin Bas/ stock.adobe.com; **P.82** *c* © es0lex/stock.adobe.com; **p.84** *tl* © fad82/stock.adobe.com, *tr* © Sean K/stock.adobe.com, *cr* © Amelia Fox/stock.adobe.com, **p.86** *tr* © absent84/stock.adobe.com; **p.88** *tl* © Zerbor/stock.adobe.com; **p.89** *tr* © ayarx oren/Shutterstock.com, *cr* © Rawpixel.com/ stock.adobe.com; **p.93** *tl* © Michele Paccione/stock.adobe.com; **p.95** *tr* © Monkey Business/ stock.adobe.com, *br* © Igor/stock.adobe.com; **p.102** *tl* © The Asahi Shimbun/Getty Images; **p.105** *cr* © Christos Georghiou/stock.adobe.com, *br* © Syda Productions/stock.adobe.com; **p.108** *r* © beysim/stock.adobe.com; **p.109** *t* © https://commons.wikimedia.org/wiki/File:Euler_diagram_ of_triangle_types.svg, https://creativecommons.org/licenses/by-sa/3.0/, **p.110** *tl* © Arkady Mazor/ Shutterstock.com, *bl* © Sashkin/stock.adobe.com, *tr* © vkara/stock.adobe.com; **p.111** *tl* © Cynthia Johnson/The LIFE Images Collection/Getty Images, *tc* Photo of Paul Erdős by George Csicsery taken in 1989 in Poznan, Poland. © Zala Films, *tr* © Bundesarchiv, Bild 183-33149-0001 / photo: Paul Turan, *bl* © Denis Topal/stock.adobe.com; **p.116** *cl* © sarymsakov.com/stock.adobe.com, *br* © Sergione/stock.adobe.com; **p.117** *bl* © Evgeny Kozhevnikov/stock.adobe.com, *cl* © IgorTravkin/ stock.adobe.com, *br* © rorf33/stock.adobe.com, *bbr* © xy/stock.adobe.com; **p.118** *b* © https:// commons.wikimedia.org/wiki/File:PrimitivePythagoreanTriplesRev08.svg, https://creativecommons. org/licenses/by-sa/3.0/; **p.122** *tr* © IgorTravkin/stock.adobe.com, *bl* © Sergione/stock.adobe. com; **p.124** *br* © larryhw/stock.adobe.com; **p.125** *tl* © rorf33/stock.adobe.com, *cl* © janevans35/ stock.adobe.com, *bl* © Rita Bateson, *br* © 123levit/stock.adobe.com; **p.126** *tl* © Malbert/stock. adobe.com, *tr* © McCollyer/Shutterstock.com; **p.127** *l* & *r* © Brad Mitchell/Alamy Stock Photo; **p.128** *tl* © giedre vaitekune/Shutterstock.com, *bl* © Rita Bateson, *tr* *l* to *r* © kreatorex/stock.

Index